Healthcare Staff Scheduling

Emerging Fuzzy
Optimization Approaches

Healthcare Staff Scheduling

Emerging Fuzzy
Optimization Approaches

Michael Mutingi
Charles Mbohwa

CRC Press
Taylor & Francis Group
Boca Raton London New York

CRC Press is an imprint of the
Taylor & Francis Group, an **informa** business

Contents

SECTION III   Research Applications

Preface

Research activities concerned with healthcare staff scheduling have often highlighted the need to consider human preferences and choices when developing decision analysis methods. Staff schedules should be flexible enough to satisfy the preferences of the healthcare worker, the expectations of the patient, and the often imprecise goals of management. Researchers naturally use heuristics to find good or near-optimal solutions. However, as healthcare workers continue to call for more individualized work schedules that incorporate their wishes and preferences, and patients continue to call for personalized healthcare attention, further research into multicriteria solution approaches to staff scheduling is imperative.

Research in metaheuristics, particularly biologically inspired metaheuristics, has been quite active in the last decades. This is evidenced by a wide spectrum of hard combinatorial problems that have been successfully handled and solved. Genetic algorithms, evolutionary algorithms, particle swarm optimization, and simulated evolution are typical examples of interactive heuristic algorithms that have been widely applied in optimization research, particularly in staff scheduling. Furthermore, research advances in fuzzy theoretic decision analysis have been widely accepted in various areas. Fuzzy theory, unlike probability theory, continues to be handy in modeling imprecision, fuzziness, vagueness, and epistemic uncertainty, which are commonplace in healthcare operations. A hybrid approach such as fuzzy heuristic optimization is therefore highly motivating.

It has become evident in the healthcare operations research community that concentrating on the use of metaheuristics alone is rather disadvantageous, given the need to further advance and widen the applications of these approaches. In real life, researchers and decision analysts need interactive, flexible, and adaptive solution approaches to staff scheduling problems. Combining the strengths of metaheuristics and fuzzy theory concepts potentially yields a more effective and efficient approach. Fuzzy set theory brings more realism into problem formulation, where some of the parameters involved are vague or imprecise, which is common in real-life problems. To model human choices, preferences, expectations, and other imprecise goals, a judicious trade-off approach is more appropriate. In this respect, the use of fuzzy theory concepts provides the decision maker with a practical method for capturing and evaluating imprecise problem situations from a trade-off perspective.

In light of the preceding discussion, it is high time for a reference text on fuzzy metaheuristic approaches that presents modern developments in this domain. The book gathers emerging research material from the authors' recent research experience in multicriteria algorithms and their applications in healthcare operations, particularly in staff scheduling. Researchers can go a step further to adapt and implement these algorithms in other areas. It is hoped that by reading this book, graduate research students, researchers, and practicing decision makers will gain an in-depth understanding of the field and obtain a clear picture for further research applications.

The research work presented comprises three sections. In Section I, the book commences by highlighting research trends and challenges in healthcare staff scheduling, together with the basic concepts of fuzzy set theory. In Section II, emerging fuzzy optimization algorithms derived from biologically inspired approaches and fuzzy theory are presented. In Section III, research applications in healthcare staff scheduling are presented, providing a practical, in-depth understanding of fuzzy metaheuristic approaches to research students, researchers, and practitioners.

Authors

Michael Mutingi is a researcher at the University of Johannesburg, South Africa, completing his PhD in engineering management at the University of Johannesburg, South Africa. He also holds a BEng and a MEng in industrial engineering, both from the National University of Science and Technology, Zimbabwe. His current research interests include healthcare operations management, biologically inspired meta-heuristic optimization, fuzzy multicriteria decision methods, and Lean management. Other areas of interest include green supply chain management, logistics management, manufacturing systems simulation, and business system dynamics.

As an author of 12 book chapters and more than 70 peer-reviewed articles in international journals and conference proceedings, Mutingi is a recipient of five best-paper awards in the area of healthcare operations management—specifically in fuzzy heuristic approaches for healthcare staff scheduling. He is the originator of the simulated metamorphosis algorithm.

Mutingi has served as a research associate at the National University of Singapore. He has also served as a lecturer in industrial engineering at the University of Botswana and the University of Science and Technology, Zimbabwe. He is currently a lecturer in industrial engineering at the Namibia University of Science and Technology, Namibia. In addition, he is a consultant in Lean and operational excellence, and an associate member of the Southern African Institute of Industrial Engineers.

Charles Mbohwa, PhD, is the vice dean of postgraduate studies, research and innovation, faculty of engineering and the built environment, University of Johannesburg, South Africa. As an established researcher and professor in sustainability engineering and operations management, his specializations include renewable energy systems, biofuel feasibility and sustainability, life cycle assessment, and healthcare operations management. He has presented at numerous conferences and published more than 150 papers in peer-reviewed journals and conferences, as well as 6 book chapters and 1 book.

Upon graduating with a BSc in mechanical engineering from the University of Zimbabwe in 1986, he served as a mechanical engineer at the National Railways of Zimbabwe. He holds an MSc in operations management and manufacturing systems from the University of Nottingham, United Kingdom, and completed his doctoral studies at Tokyo Metropolitan Institute of Technology, Japan.

Professor Mbohwa was a Fulbright scholar visiting the Supply Chain and Logistics Institute at the School of Industrial and Systems Engineering, Georgia Institute of Technology. He has been a collaborator with the United Nations Environment Program and a visiting exchange professor at Universidade Tecnologica Federal do Parana in Brazil. He has also visited many countries on research and training engagements including the United Kingdom, Japan, Germany, France, the United States, Brazil, Sweden, Ghana, Nigeria, Kenya, Tanzania, Malawi, Mauritius, Austria, the Netherlands, Uganda, Namibia, and Australia. He is a fellow of the Zimbabwe Institution of Engineers and a registered engineer with the Engineering Council of Zimbabwe.

Section I

Introduction

1 Introduction

1.1 BACKGROUND

Healthcare staff scheduling plays a pivotal role in making decisions that contain and control operations costs, particularly labor costs (Ernst et al., 2004; Mutingi and Mbohwa, 2013). As labor costs continue to be a major concern to most decision makers in the healthcare sector, efficient coordination of healthcare activities is crucial. In this regard, developing new and enhanced decision support tools is highly imperative. However, scheduling healthcare staff is a complex problem area.

Oftentimes, the scheduling decision process is affected by imprecise information on patient expectations, service demand, staff preferences, and management goals (Topaloglu and Selim, 2010; Mutingi and Mbohwa, 2013). The decision maker utilizes the perceived information to construct efficient and cost-effective work schedules. However, in the presence of fuzzy variables, the impact of the decisions taken is unknown at the planning or scheduling stage (Mutingi and Mbohwa, 2013). Therefore, in the real world, healthcare staff scheduling should consider fuzziness imposed by four main factors: (1) imprecise management goals, (2) imprecise staff preferences, (3) imprecise patients' preferences, and (4) conflicting goals and preferences. Thus, goals regarding service levels, staff schedule fairness, and quality of service are often expressed qualitatively, rather than quantitatively (Mutingi and Mbohwa, 2013). Clearly, vagueness or imprecision arising from the use of natural language in defining model variables is nonprobabilistic in nature (Bezdek, 1993; Mutingi and Mbohwa, 2014). For example, when constructing work schedules for nursing staff, it is difficult to incorporate the various vague wishes and preferences of the staff. Consequently, the use of probability theory alone is not adequate. As such, the decision maker has to rely on the available information and expert knowledge to formulate effective staff scheduling decisions. Hence, it is important to develop fuzzy heuristic approaches that can address uncertainties in healthcare staff scheduling. The core areas of this research are staff scheduling, staff routing and scheduling, and task assignment.

1.2 HEALTHCARE STAFF SCHEDULING

Healthcare staff problems are presented in different forms in the literature. Numerous empirical and hypothetical problem instances have been reported (Cheang et al., 2003; Ernst et al., 2004; Mutingi and Mbohwa, 2013). In this research, five main areas of healthcare staff scheduling are presented, namely:

1. Nurse scheduling
2. Nurse rerostering
3. Physician scheduling
4. Homecare scheduling
5. Care task assignment

3

TABLE 1.1

The Five Selected Areas of Healthcare Staff Scheduling

No.	Scheduling Problem	Brief Description	Selected References
1	Nurse scheduling	This involves construction of work schedules in the short to medium term.	Topaloglu (2006); Topaloglu and Selim (2010); Cheang et al. (2003)
2	Nurse rerostering	This is the process of reconstructing a nursing schedule, beginning from the reported first day of absences up to the last day of the roster.	Maenhout and Vanhoucke (2011); Moz and Pato (2007)
3	Physician scheduling	This is the process of scheduling physicians in an emergency department so that various unexpected injuries and illnesses are adequately and urgently attended to.	Lo and Lin (2011); Puente et al. (2009)
4	Homecare scheduling	This entails staff routing and scheduling concerned with construction of coordinated staff visits to patients in a homecare setting.	Bertels and Fahle (2006); Mutingi and Mbohwa (2014); Akjiratikarl, Yenradee, and Drake (2007)
5	Care task assignment	This involves short-term or daily assignment of tasks to available nursing staff.	Cheng et al. (2007); Lin and Yeh (2007)

Table 1.1 provides a brief description of healthcare staff scheduling problems, together with pertinent selected references.

A description of the specific problem areas is presented in the next sections.

1.2.1 Nurse Scheduling

Nursing rostering or scheduling is the process of constructing work schedules so as to satisfy healthcare service needs, patient preferences, employee preferences, availability of resources, organizational goals, and other workplace regulations (Cheang et al., 2003; Ernst et al., 2004). Real-world staff scheduling is associated with inherent uncertainties due to vagueness of information on target management objectives and personnel preferences. Burke et al. (2004) emphasize the need to consider fuzzy modeling in nurse scheduling in order to deal with the inherent uncertainties. Bard and Purnomo (2005) argued that preference-based scheduling increases job satisfaction. Preferences are measured in the form of requests for specific shift patterns, duties, tasks, or number of working hours. While this feature has the potential to improve schedule personalization, and hence job satisfaction, this will definitely increase the complexity of the scheduling problem (Topaloglu, 2006; Topaloglu and Selim, 2010).

The appropriateness of nurse schedules has a great impact on the quality of healthcare, the development of healthcare budgets, and various operations of the healthcare organization. However, it is difficult to incorporate staff preferences; the decision maker has to take into account the wishes and preferences of the patients and the nursing staff, not forgetting the management goals, which are often expressed imprecisely (Topaloglu and Selim, 2010; Mutingi and Mbohwa, 2013, 2014).

The identified complicating features inherent in the nursing staff scheduling problem are as follows:

1. The presence of vague staff wishes and preferences concerning fair workloads, shift equity, and workmate congenialities
2. The presence of imprecise management goals and choices, which are difficult to incorporate in conventional solution methods
3. The conflicting nature of management goals and choices, staff preferences, and expectations
4. The presence of multiple conflicting objectives and constraints, which makes the problem difficult to solve

In practice, it is difficult to satisfy imprecise conflicting management goals, worker preferences, and constraints. The staff scheduling problem is hard, highly combinatorial, and constrained, demanding a lot of computation time, with little solution improvement. On the other hand, decision makers prefer interactive, fast solution methods that provide a population of alternative solutions, rather than prescribe a single solution. Research should seek to cover these voids by developing heuristic, interactive, multicriteria approaches. It is therefore necessary to provide flexible frameworks that address imprecise management objectives and worker preferences and choices.

1.2.2 NURSE REROSTERING

In most hospitals, it happens every so often that after a nurse roster has been assigned, nurses announce that they will be unable to show up for one or more of their shifts. In the absence of a reserve pool of nurses to substitute, the roster has to be rebuilt to accommodate the unforeseen changes (Maenhout and Vanhoucke, 2011). Therefore, the nurse rerostering problem is concerned with reconstruction of an existing roster spanning the first day of absences up to the last day of the roster. However, the new roster is also subject to all the constraints pertaining to the original one. In addition, the new roster must not allocate any shifts to the nurse that will be absent. As expected, nurses and management would wish to have a new roster that is as similar to the original as possible, minimizing the number of shift swaps in the nurse roster. In this regard, the aim is to build a new feasible roster considering the imprecise wishes and preference of nurses, as well as management goals.

In nurse rerostering, it is important to note that the resulting nursing hours in each day of the new roster must meet the total minimum requirement (Moz and Pato, 2007). This ensures that the level of healthcare satisfies the care standards in the hospital units.

1.2.3 PHYSICIAN SCHEDULING

In a hospital setting, an emergency room or department is a medical treatment facility that provides 24-hour specialist-led emergency care. Patients of the emergency department almost always come with various unexpected injuries and illnesses that require urgent treatment (Lo and Lin, 2011). As a result, the nature of work in the emergency room department is unique in that it is highly unpredictable and stressful to physicians and other associated medical staff (Lin and Yeh, 2007). Physician scheduling is a challenging problem that has recently attracted the attention of most decision makers in hospitals. However, there are some similarities between physician scheduling and nurse rostering.

Like nurse rostering, physician scheduling seeks to construct shift assignments that maximize staff satisfaction, patient satisfaction, and management goals, as well as meet hospital regulations and policies. Thus, both problems are time consuming and complex. Furthermore, nurse rostering and physician scheduling are multicriteria decision problems that demand efficient and effective decision support tools that can accommodate the choices, wishes, and preferences of patients and staff, as well as of the decision maker.

1.2.4 HOMECARE NURSE SCHEDULING

Homecare service is often confronted with complex decisions concerned with routing nursing staff who should visit patients in their homes (Bertels and Fahle, 2006; Bard, Shao, and Wang, 2013). The main decision consists in constructing coordinated nurse visits to patients for treatment or care activities during a prespecified time window (Akjiratikarl, Yenradee, and Drake, 2007; Mutingi and Mbohwa, 2014). The decision is motivated by ever-increasing home healthcare needs, pressure for high-quality healthcare services, and the desire to contain healthcare labor costs. The constructed schedules are expected to satisfy, as much as possible, the expectations and wishes of caregivers, patients, and management. Home healthcare nurse scheduling has recently attracted the attention of decisions makers in the healthcare sector. It is essential to develop efficient models that can handle the homecare nurse scheduling problem from a multicriteria point of view, considering patient preferences, shift equity, individual staff preferences, and management goals. However, these expectations are often imprecise or fuzzy, suggesting the use of fuzzy multicriteria optimization approaches.

In view of this, homecare nurse scheduling is characterized by three major complicating features: (1) the presence of vague staff wishes and preferences such as fair workloads, shift equity, choice of specific shift types, and workmate congenialities; (2) the presence of vague patient expectations and preferences in regard to the choice of visit time windows, and the choice of specific nursing staff to visit them; and (3) the presence of imprecise conflicting management goals and choices, which are difficult to quantify and incorporate in a conventional solution approach. Therefore, developing new or improved efficient multicriteria optimization methods is imperative.

1.2.5 CARE TASK ASSIGNMENT

Care task assignment in hospitals is concerned with allocation of nursing activities for assistance with meals, instillation of drops, and preparation of infusions

(Paulussen et al., 2003; Cheng et al., 2007). Hard and soft constraints pertaining to nurse capacity limitations, nurse preferences, relationships between care tasks, and management goals have to be observed. While nurses expect fair task assignment, patient expectations and management goals must be satisfied as much as possible. The presence of imprecise conflicting preferences and goals makes task assignment complex.

In practice, care task assignment is often done manually using spreadsheets, which may be extremely time consuming. Conventional dispatching rules, such as earliest due date, slack, extended slack, and first in/first out, may be used (Cheng et al., 2007). However, the rules have a rigid structure; they disallow the use of multiple criteria. In summary, healthcare task assignment is a challenging problem characterized by (1) vague conflicting staff preferences, (2) imprecise patient preferences and care task time windows, (3) imprecise management goals and choices that are difficult to model and quantify, and (4) the need to handle multiple decision criteria.

1.3 RESEARCH CHALLENGES

Interesting research challenges can be derived from the previously mentioned healthcare scheduling problems. Some of the key challenges and research directions are presented next.

1.3.1 NEED FOR PRACTICAL MULTICRITERIA REASONING

Multicriteria decision making is a recent decision approach that is applicable to a wide range of real-world problems (Marquis et al., 2007; Vasant et al., 2007; Elamvazuthi et al., 2009; Madronero, Peidro, and Vasant, 2010; Villacorta, Masegosa, and Lamata, 2013). Healthcare staff scheduling usually presents a range of objectives and restrictions, which may be conflicting (Burke et al., 2004). For example, satisfying staff preferences may conflict with satisfying shift demand. A handful of cases of multicriteria reasoning have been investigated in literature (Berrada, Ferland, and Michelon, 1996; Jaszkiewicz, 1997; Burke et al., 2002). Decision support systems in healthcare systems should reflect the inherent multiple criteria in staff scheduling. In this view, further research in this gray area is highly promising.

1.3.2 NEED FOR ADAPTABILITY AND FLEXIBILITY

One major focus area in multicriteria decision making is qualitative flexible soft computing (Vasant, 2006; Villacorta et al., 2014). This is because decision makers in healthcare usually expect flexible heuristic scheduling approaches that can adapt to specific problem scenarios. Heuristic algorithms are expected to tackle practical issues in the most flexible way possible, which closed-form models may not be able to address adequately (Vasant, 2011). To that effect, fuzzy modeling techniques can be used to infuse flexibility and adaptability into heuristic algorithms. Fuzzy modeling allows decision makers' choices to be modeled into one or more of the algorithmic functions.

1.3.3 NEED FOR INTERACTIVE DECISION SUPPORT

Practicing decision makers in healthcare prefer to use decision support tools such as heuristic algorithms that can provide a set of good alternative solutions, rather than prescribe a single solution. This gives room for further adjustments on parameter inputs and other decision choices. Furthermore, the selection of the final solution should rest upon the decision maker, who may need to undertake other practical considerations that may not have been incorporated directly into the structure of the decision support tool.

1.3.4 NEED FOR FUZZY THEORETIC APPROACHES

To address uncertainties in real-world healthcare staff scheduling, fuzzy set theory is a crucial addition to decision support tools. For instance, it can be used to address uncertainties involving the human factor with all its vagueness of desired goals, perceptions, ambiguity, and subjectivity. Fuzzy models can utilize fuzzy memberships to represent similarities of objects with imprecisely defined properties. Desired goals and preferences, such as "close to 36 hours per week per worker," "preferably day shifts," and "not more than six shifts but much more than two shifts per week," can be modeled using fuzzy memberships. Fuzzy logic derives its importance from the fact that human reasoning and common sense tend to be approximate in nature. Classical crisp set theory is limited in the presence subjective human perceptions. Therefore, the challenge is the application of fuzzy set theory concepts from a multi-criteria viewpoint.

1.4 FUZZY SET THEORY: PRELIMINARIES

Fuzzy set theory, formally introduced by Zadeh (1965, 1978), is a flexible tool for solving real-world decision problems involving fuzzy or imprecise aspects. Human factors play a vital role in the overall behavior of most real-world systems. Much of the decision making takes place in a fuzzy environment where the desired goals and the consequences of decisions are not precisely known. Fuzzy theory helps the decision maker not only to evaluate existing alternatives, but also to develop new alternatives. Its techniques have been applied in a number of areas such as optimization theory, control systems, artificial intelligence, and human behavior (Inuiguchi and Ramik, 2000; Vasant, 2006, 2011). Therefore, fuzzy theory is a potential tool for solving management decision problems in a fuzzy environment.

Zadeh (1965) coined the term "fuzzy" to define uncertainty in the nonstochastic sense rather than random variables. A fuzzy set is a class of objects in which there is no sharp boundary between objects that belong to a class and those that do not. Contrary to the two-valued conventional Boolean logic, fuzzy sets deal with degrees of membership and degrees of truth, acknowledging that things can be partly true and partly false at the same time. In this view, fuzzy logic is a superset of Boolean logic that has been extended to handle partial truth-values between completely true and completely false (Bezdek, 1993). From a mathematical point of view, fuzzy sets are a generalization of classical sets. Formally, a more precise definition can be given as follows:

Definition 1.1: Fuzzy Set

Let $X = \{x\}$ be a collection of points (objects), called the universe, whose generic elements are denoted by x. Then, a fuzzy set A in X is a set of ordered pairs defined in terms of its characteristic function $\mu_A(x)$,

$$A = \{(x, \mu_A(x))\}, \quad x \in X \tag{1.1}$$

where $\mu_A(x)$ represents the grade of membership of x in A, and $\mu_A: X \rightarrow M$ is a function that maps X to a membership space M in the interval [0,1], with 0 and 1 representing the lowest and highest grade of membership, respectively. ∎

Here, the characteristic function is also called a possibility distribution. Thus, a high value of $\mu_A(x)$ implies that it is very likely for x to be in A. It follows from the preceding definition that in classical nonfuzzy set theory, a point x belongs to set A if and only if $\mu_A(x) = 1$ and does not belong to the set if $\mu_A(x) = 0$. In most practical fuzzy environments, $\mu_A(x)$ has to be estimated from partial data over a finite set of sample n points $x_1, \ldots, x_n$.

1.5 OUTLINE OF THE BOOK

This book focuses on research on healthcare staff scheduling and the application of modern fuzzy heuristic optimization methods. Healthcare operations, both in hospital and in home healthcare settings, are overwhelmed with complex fuzzy features that impose difficulties in the construction of work schedules. Scheduling decisions are generally impacted by imprecise management goals, staff preferences, and patients' preferences. As a result, the decision maker has to rely on the use of expert intuition and knowledge. This book brings together recent research on fuzzy multicriteria optimization approaches that incorporate concepts of fuzzy sets and metaheuristic algorithms for staff scheduling in healthcare settings.

The book comprises three sections. The first introduces the pertinent research challenges in the healthcare sector. In this regard, Chapter 1 highlights recent research challenges and trends in healthcare staff scheduling, pointing out the relevance of using fuzzy theory concepts and fuzzy evaluation techniques.

Section II (Chapters 2–5) presents emerging metaheuristic approaches and the use of fuzzy set theory concepts to address complex multicriteria decisions. Chapter 2 focuses on developing a novel fuzzy simulated metamorphosis (FSM) algorithm deriving from the biological concepts of metamorphosis in insects. A range of potential application areas is then presented.

Chapter 3 proposes an enhanced fuzzy simulated evolution (FSE) algorithm that uses fuzzy multicriteria techniques to model and evaluate alternative solutions. Typical application areas of the proposed algorithm are outlined.

Chapter 4 proposes a unique fuzzy grouping genetic algorithm (FGGA), which uses unique genetic operators that are adaptable to grouping problems. A range of key grouping problems is then presented. Scheduling problems with a characteristic grouping structure lend themselves to the proposed algorithm.

Chapter 5 presents an enhanced fuzzy grouping particle swarm optimization (FGPSO) algorithm, developed from the basic particle swarm optimization (PSO) algorithm. By adding grouping techniques and fuzzy evaluation methods to the basic PSO algorithm, a unique algorithm is developed. The method is applicable to notable grouping problems in the healthcare sector.

Section III (Chapters 6–11) focuses on research applications in healthcare staff scheduling, providing researchers and practitioners with a practical, in-depth understanding of fuzzy metaheuristic approaches. To model the fuzzy features of staff scheduling problems, conflicting fuzzy management goals, staff preferences, and patient expectations are expressed and evaluated as fuzzy membership functions. The research applications of the approaches fall into three categories. The first is concerned with constructing and reconstructing high-quality staff schedules spanning a period of 1 week or more. The second category focuses on developing high-quality staff schedules in a home healthcare setting. The third category deals with assignment of tasks in the short term—that is, on a daily basis. The overall objective is to optimize costs, and to prevent violation of staff and patient preferences, subject to time and staff capacity constraints.

Chapter 6 focuses on the nurse scheduling or nurse rostering problem with the application of the fuzzy simulated metamorphosis algorithm. The chapter seeks to (1) present the nurse scheduling problem in a hospital setting, (2) present a multicriteria fuzzy evolutionary algorithm deriving from biological metamorphosis, and (3) apply the algorithm to nurse scheduling problems, demonstrating its efficiency and effectiveness.

Chapter 7 presents the nurse rerostering problem, a unique but common problem in the healthcare sector. The problem is modeled using a multicriteria fuzzy simulated evolution algorithm. The objectives of the chapter are to (1) present the nurse rerostering problem, (2) apply the fuzzy simulated evolution algorithm, and (3) present illustrative computational experiments, so as to demonstrate the effectiveness of the method.

Chapter 8 models the physician scheduling problem using a multicriteria fuzzy particle swarm optimization method. As in the fuzzy grouping particle swarm optimization, fuzzy multicriteria techniques are used to model imprecise preferences and choices of the decision maker, the patients, and the physicians.

Chapter 9 presents a multicriteria approach to homecare staff scheduling with the application of a fuzzy grouping genetic algorithm. The research work follows through three objectives, that is, (1) to describe the homecare staff scheduling problem with time windows, (2) to propose a fuzzy grouping genetic algorithm approach for solving the problem, and (3) to carry out illustrative computational experiments using the proposed approach.

Chapter 10 considers the care task assignment problem from a multicriteria viewpoint, and solves the problem based on fuzzy grouping genetic algorithm. The specific research objectives are to (1) describe the care task assignment problem, (2) propose a fuzzy grouping particle swarm optimization, and (3) provide illustrative examples, demonstrating the effectiveness of the algorithm.

Chapter 11 summarizes the work covered in this book and derives conclusions from the research work, providing further research avenues related to the current research area.

The Appendix details fuzzy set theory concepts.

The emerging approaches proposed in this book are beneficial in three main ways: (1) they incorporate fuzzy preferences and decision makers' choices, giving more realism to the approaches; (2) they are flexible and adaptive to problem situations, providing room for interactive decision support for decision makers; and (3) they provide reliable solutions within reasonable computation times. Therefore, these approaches are very useful to researchers, academicians, and practicing decision makers concerned with healthcare staff scheduling.

REFERENCES

Akjiratikarl, C., Yenradee, P. and Drake, P. R. 2007. PSO-based algorithm for home care worker scheduling in the UK. *Computers & Industrial Engineering* 53: 559–583.

Bard, J. F. and Purnomo, H. W. 2005. Preference scheduling for nurses using column generation. *European Journal of Operational Research* 164 (2): 510–534.

Bard, J. F., Shao, Y. and Wang, H. 2013. Weekly scheduling models for travelling therapists. *Socio-Economic Planning Sciences* 47 (3): 191–204.

Berrada, I., Ferland, J. A. and Michelon, P. 1996. A multi-objective approach to nurse scheduling with both hard and soft constraints. *Socio-Economic Planning Science* 30 (20): 183–193.

Bertels, S. and Fahle, T. 2006. A hybrid setup for a hybrid scenario: Combining heuristics for the home health care problem. *Computers & Operations Research* 33 (10): 2866–2890.

Bezdek, J. C. 1993. Editorial: Fuzzy models—What are they and why? *IEEE Transactions on Fuzzy Systems* 1 (1): 1–6.

Burke, E., De Causmaecker, P., Vanden Berghe, G. and Landeghem, H. 2004. The state of the art of nurse rostering. *Journal of Scheduling* 7: 441–499.

Burke, E. K., De Causmaecker, P., Petrovic, S. and Vanden Berghe, G. 2002. A multicriteria metaheuristic approach to nurse rostering. In *Proceedings of Congress on Evolutionary Computation*, CEC 2002, Honolulu, IEEE Press, 1197–1202.

Cheang, B., Li, H., Lim, A. and Rodrigues, B. 2003. Nurse rostering problems—A bibliographic survey. *European Journal of Operational Research* 151: 447–460.

Cheng, M., Ozaku, H. I., Kuwahara, N., Kogure, K. and Ota, J. 2007. Nursing care scheduling problem: Analysis of staffing levels. *Proceedings of the 2007 IEEE International Conference on Robotics and Biomimetics*, December 15–18, 2007, Sanya, China, 1: 1715–1719.

Elamvazuthi, I., Ganesan, T., Vasant, P. and Webb, J. F. 2009. Application of a fuzzy programming technique to production planning in the textile industry. *International Journal of Computer Science and Information Security* 6 (3): 238–243.

Ernst, A. T., Jiang, H., Krishnamoorthy, M. and Sier, D. 2004. Staff scheduling and rostering: A review of applications, methods and models. *European Journal of Operational Research* 153: 3–27.

Inuiguchi, M. and Ramik, J. 2000. Possibilistic linear programming: A brief review of fuzzy mathematical programming and a comparison with stochastic programming in portfolio selection problem. *Fuzzy Sets and Systems* 111 (1): 3–28.

Jaszkiewicz, A. 1997. A metaheuristic approach to multiple objective nurse scheduling. *Foundations of Computing and Decision Sciences* 22 (3): 169–184.

Lin, W. S. and Yeh, J. Y. 2007. Using simulation technique and genetic algorithm to improve the quality care of a hospital emergency department. *Expert Systems with Applications* 32: 1073–1083.

Lo, C.-C. and Lin, T.-H. 2011. A particle swarm optimization approach for physician scheduling in a hospital emergency department. *IEEE Seventh International Conference on Natural Computation* 1929–1933.

Madronero, M. D., Peidro, D. and Vasant, P. 2010. Vendor selection problem by using an interactive fuzzy multi-objective approach with modified s-curve membership functions. *Computers and Mathematics with Applications* 60: 1038–1048.

Maenhout, B. and Vanhoucke, M. 2011. An evolutionary approach for the nurse rerostering problem. *Computers & Operations Research* 38: 1400–1411.

Marquis, J., Fowler, J. W., Gel, E., Köksalan, M., Korhonen, P. and Wallenius, J. 2007. Interactive evolutionary multi-criteria scheduling. *Proceedings of the 3rd Multidisciplinary International Conference on Scheduling: Theory and Applications*, 2007, 591–594.

Moz, M. and Pato, M. V. 2007. A genetic algorithm approach to a nurse rerostering problem. *Computers & Operations Research* 34: 667–691.

Mutingi, M. and Mbohwa, C. 2013. Fuzzy modeling for manpower scheduling. In *Exploring innovative and successful applications of soft computing*, eds. Carlos Cruz-Corona, M. S. García-Cascales, María Teresa Lamata, Antonio David Masegosa, José Luis Verdegay and Pablo J. Villacorta. IGI-Global, USA, 138–160.

Mutingi, M. and Mbohwa, C. 2014. Multi-objective homecare worker scheduling: A fuzzy simulated evolution algorithm approach. *Transactions on Healthcare Systems Engineering* 4: 1–8.

Paulussen, T. O., Jennings, N. R., Decker, K. S. and Heinzl, A. 2003. Distributed patient scheduling in hospitals. *Proceedings of the 18th International Joint Conference on Artificial Intelligence*, 1224–1229.

Puente, J., Gomez, A., Fernandez, I. and Priore, P. 2009. Medical doctor rostering problem in a hospital emergency department by means of genetic algorithms. *Computers & Industrial Engineering* 56: 1232–1242.

Topaloglu, S. 2006. A multi-objective programming model for scheduling emergency medicine residents. *Computers & Industrial Engineering* 51: 375–388.

Topaloglu, S. and Selim, S. 2010. Nurse scheduling using fuzzy modeling approach. *Fuzzy Sets and Systems* 161: 1543–1563.

Vasant, P. 2006. Fuzzy production planning and its application to decision making. *Journal of Intelligent Manufacturing* 17 (1): 5–12.

Vasant, P. 2011. Hybrid MADS and GA techniques for industrial production systems. *Archives of Control Sciences* 21 (3): 227–240.

Vasant, P., Bhattacharya, A., Sarkar, B. and Mukherjee, S. K. 2007. Detection of level of satisfaction and fuzziness patterns for MCDM model with modified flexible S-curve MF. *Applied Soft Computing Journal* 7 (3): 1044–1054.

Villacorta, P. J., Masegosa, A. D., Castellanos, D. and Lamata, M. T. 2014. A new fuzzy linguistic approach to qualitative cross impact analysis. *Applied Soft Computing* 24: 19–30.

Villacorta, P. J., Masegosa, A. D. and Lamata, M. T. 2013. Fuzzy linguistic multi-criteria morphological analysis in scenario planning. *IEEE IFSA World Congress and NAFIPS Annual Meeting (IFSA/NAFIPS)*, June 2013, 777–782.

Zadeh, L. A. 1965. Fuzzy sets. *Information and Control* 8: 338–353.

Zadeh, L. A. 1978. Fuzzy sets as a basis for a theory of possibility. *Fuzzy Sets and Systems* 1: 3–28.

Section II

Emerging Fuzzy Optimization Approaches

2 Fuzzy Simulated Metamorphosis Algorithm

2.1 INTRODUCTION

Addressing multicriteria decision problems in a fuzzy environment characterized by conflicting goals requires fast and flexible decision support tools that are easily adaptable to different problem situations. More often than not, decision makers desire to incorporate imprecise expert choices, judgments, preferences, and intuitions from relevant sources into their decision-making processes. For instance, in nurse scheduling, a decision maker may want to incorporate his or her choices, expert opinion from management, nurse preferences regarding shift allocation, and patient preferences (Mutingi and Mbohwa, 2014). In such problem contexts, decision makers desire to use judicious approaches to find a cautious trade-off between conflicting goals. This is common in a wide range of real-world problems such as vehicle routing (Shaffer, 1991; Tarantilis, Kiranoudis, and Vassiliadis, 2004), staff scheduling (Ernst et al., 2004), nurse scheduling (Topaloglu and Selim, 2010), and job shop scheduling (Sakawa and Kubota, 2000). Addressing ambiguity, imprecision, and uncertainties of conflicting management goals is highly desirable in such problem settings.

Recently, research has shown that biologically inspired metaheuristic algorithms are quite efficient and effective in solving real-world optimization problems. A review of metaheuristics is given in Osman and Laporte (1996). These algorithms include genetic algorithms, simulated evolution algorithms, ant colony optimization, and particle swarm optimization. However, to tackle complex fuzzy multicriteria, more research in this direction is essential.

In view of the wide range of real-world problem situations that require fuzzy multicriteria decision support, developing efficient interactive fuzzy metaheuristics is highly imperative. In particular, such metaheuristic approaches should necessarily be able to incorporate the decision maker's choices, intuitions and expert judgments into their iterative procedures. The research purpose of this chapter is to present a fuzzy simulated metamorphosis algorithm deriving from the natural biological process of metamorphosis common in insects (Truman and Riddiford, 2002). As such, the objectives of the research are as follows:

1. To present the basic simulated metamorphosis algorithm
2. To present a fuzzy simulated metamorphosis algorithm
3. To discuss the application areas of the fuzzy simulated metamorphosis algorithm

The rest of the chapter is structured as follows. The next section presents the concepts of metamorphosis evolution followed by the basic metamorphosis algorithm. A fuzzy simulated metamorphosis algorithm is then presented from a more general point of view. This is followed by a comparative analysis between the fuzzy simulated metamorphosis algorithm and other competitive algorithms. An outline of potential application areas is provided. Finally, a summary of the chapter is presented.

2.2 BIOLOGICAL METAMORPHOSIS EVOLUTION: THEORY AND CONCEPTS

Metamorphosis is a biological evolutionary process common in insects such as butterflies (Truman and Riddiford, 2002). As illustrated in Figure 2.1, the process begins with an egg that hatches into an instar larva (or instar). Subsequently, the first instar transforms into several instars, then into a pupa, and finally into the adult insect (Tufte, 2011). The process is uniquely characterized with hormone-controlled growth with radical evolution and maturation.

2.2.1 METAMORPHOSIS EVOLUTION

When an insect grows and develops, it must periodically shed its rigid exoskeleton in a process called molting. The insect grows a new, loose exoskeleton that provides the insect with room for more growth (Truman and Riddiford, 2002). The insect transforms in body structure as it molts from a juvenile to an adult form, a process called metamorphosis.

The concept of metamorphosis refers to the change of physical form, structure, or substance—a marked and more or less abrupt developmental change in the form or structure of an animal (such as a butterfly or a frog) occurring subsequent to hatching or birth (Tufte, 2011). A species changes body shape and structure at a particular point in its life cycle, such as when a tadpole turns into a frog. Sometimes, in locusts for example, the juvenile form is quite similar to the adult one. In others, they are

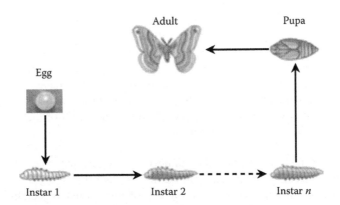

FIGURE 2.1 Biological metamorphosis evolution.

radically different and unrecognizable as the same species. The different forms may even entail a completely new lifestyle or habitat, such as when a ground-bound, leaf-eating caterpillar turns into a long-distance flying, nectar-eating butterfly.

2.2.2 HORMONAL CONTROL

Insect molting and development are controlled by several hormones (Tufte, 2011). The hormones trigger the insect to shed its exoskeleton and, at the same time, grow from smaller juvenile forms (e.g., a young caterpillar) to larger adult forms (e.g., a winged moth). The hormone that causes an insect to molt is called ecdysone. The hormone, in combination with another, called the juvenile hormone, also determines whether the insect will undergo metamorphosis.

2.3 SIMULATED METAMORPHOSIS ALGORITHM

The simulated metamorphosis (SM) algorithm, originally developed by Mutingi and Mbohwa (2014), is an evolutionary metaheuristic approach inspired by natural biological metamorphoses common in many insect species. There are three basic phases in the SM algorithm: *initialization*, *growth*, and *maturation*. Each of these phases has specific operators, as outlined in Figure 2.2.

In the *initialization phase*, an initial feasible solution is created as a seed for the evolutionary algorithm. The initial candidate solution constituent elements e_i ($i = 1,2,...$) are evaluated and transformed over two iterative loops—namely, growth and maturation. The *growth phase* comprises evaluation, transformation, and regeneration operators. Evaluation measures the relevant quality of the candidate solution in terms of the fitness of each element in the solution. At each iteration t, the transformation operator selects and transforms weak elements at a probability p_t, a parameter that emulates the inhibition/juvenile hormone, controlling the growth rate of the solution. Elements with low fitness are subjected to growth and compared with the rejected ones, while preserving better ones. A predetermined number of rejected elements are kept in a set Q for future use in regeneration. The regeneration operator

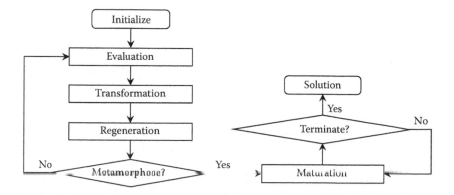

FIGURE 2.2 Simulated metamorphosis algorithm.

repairs incomplete or infeasible solutions created by the transformation operator. Elements in the reject list Q are used as food for the repair mechanism. The growth loop continues until a certain level of the juvenile hormone is reached. The *maturation phase* loops intensification and postprocessing to bring the candidate solution to maturity. While intensification ensures a complete search of an improved solution in the neighborhood of the current solution, postprocessing allows the user to interactively make expert changes to the candidate solution, and to rerun the intensification operator. As such, expert knowledge and intuition are incorporated into the solution procedure.

2.4 FUZZY SIMULATED METAMORPHOSIS ALGORITHM

Fuzzy simulated metamorphosis (FSM) is a multicriteria decision approach based on the concepts of biological metamorphosis and fuzzy evaluation techniques. The general procedure of the algorithm follows through initialization, growth, and maturation phases. The algorithm is motivated by several fuzzy multicriteria decision problems in the operations research and operations management community, such as vehicle routing problems (Tarantilis et al., 2004), nurse scheduling (Jan, Yamamoto, and Ohuchi, 2000; Inoue et al., 2003), and task assignment (Cheng et al., 2007). Associated with multiple imprecise conflicting goals, fuzzy decision problems need interactive decision support tools that can incorporate the choices, intuitions, and expert judgments of the decision maker (Topaloglu and Selim, 2010). As a fuzzy multicriteria heuristic approach, FSM seeks to bridge this gap.

2.4.1 INITIALIZATION

In the initialization phase, an initial solution is generated as a seed based on a specific heuristic guided by problem constraints. Alternatively, a decision maker can enter a user-generated solution as a seed. The aim is to obtain a feasible initial solution. The initial candidate solution, s_1, consists of constituent elements e_i $(i = 1,...,I)$, where I is the number of constituent elements in the candidate solution. Following the initialization phase, the algorithm subsequently loops through growth and maturation phases.

2.4.2 GROWTH

The growth loop consists of three operators—that is, fuzzy evaluation, transformation, and regeneration.

2.4.2.1 Fuzzy Evaluation

The nature of the evaluation function plays a key role in the success of the evaluation operator and the efficiency of the overall algorithm. First, the evaluation function should measure the relevant quality of a candidate solution. Second, the function should accurately capture the key characteristics of the problem, particularly the imprecise, conflicting, and multiobjective goals and constraints. Third, the function should be easy to compute.

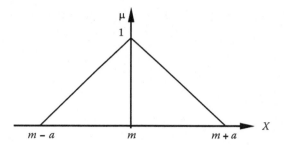

FIGURE 2.3   Symmetrical triangular membership function.

The evaluation function F_t, at iteration t, is a function of weighted normalized objective functions denoted by μ_h ($h = 1,...,n$), where n is the number of constituent objective functions. In this research, a fuzzy multifactor evaluation method is used in the form

$$F_t(s_t) = \sum_h w_h \mu_h(s_t)$$ (2.1)

where s_t is the current solution at iteration t and w_h is the weight of the function μ_h. Here, the use of the max–min operator is discouraged in order to avoid possible loss of vital information.

In practice, linear functions such as the triangular membership function are effective. The symmetrical fuzzy triangular membership function is adopted in most applications, as shown in Figure 2.3.

According to the triangular membership function, the satisfaction level is represented by a fuzzy number $A<m,a>$, where m represents the center of the fuzzy parameter with width a. Thus, the corresponding membership function is

$$\mu_A(x) = \begin{cases} 1 - \dfrac{|m - x|}{a} & \text{If } m - a \leq x \leq m + a \\ 0 & \text{If otherwise} \end{cases}$$ (2.2)

2.4.2.2 Transformation

The growth process is achieved through selection and transformation operators. Selection determines whether an element e_i of the candidate solution s_t should be retained for the next iteration, or selected for a transformation operation. The fitness η_i of element e_i ($i = 1,...,I$) is compared with probability $p_t \in [0,1]$, generated at each iteration t. That is, if $\eta_i \leq p_t$, then e_i is transformed; otherwise, it is retained intact into the next iteration. It follows from this analysis that elements with low fitness are subjected to growth. To guarantee convergence to maturity, emulating the inhibition or juvenile hormone in the biological metamorphosis evolution, the magnitude of p_t decreases over time, following a decay function of the form

$$p_t = p_0 e^{-at/T} \tag{2.3}$$

where

$p_0 \in [0,1]$ is a randomly generated number at time t
T is the maximum number of iterations
a is an adjustment factor

To avoid compete loss of better performing elements, new elements are compared with the rejected ones; the best are kept. A predetermined number of rejected elements are kept in the reject list Q for future use in regeneration. Emulating biological metamorphosis, rejected elements are used as food in the regeneration process.

2.4.2.3 Regeneration

The regeneration operator contains a repair mechanism that checks the feasibility of the candidate solution and repairs all its incomplete elements. The repair mechanism should be derived from domain-specific heuristics, considering problem constraints. After regeneration, the candidate solution is tested for readiness for transition to the maturation phase, based on the dissatisfaction level (juvenile hormone) m_t at iteration t

$$m_t = 1 - (\mu_1 \wedge \mu_2 \wedge \ldots \wedge \mu_n) \tag{2.4}$$

Here, $\mu_1, \ldots, \mu_n$ represent the satisfaction level of the respective objective functions, and "$\wedge$" is the min operator. If the dissatisfaction level is not acceptable, the growth loop repeats, until a predefined level m_0 is reached. However, if there is no significant change in m_t after a predetermined number of trials, the algorithm proceeds to the maturation phase.

2.4.3 MATURATION

The purpose of the maturation phase is to refine the candidate solution to maturity, preferably obtaining the best solution. The maturation loop consists of intensification and postprocessing operators.

2.4.3.1 Intensification

The purpose of the intensification operator is to perform a thorough search of better solutions in the neighborhood of the current solution. This helps to improve the current solution further. It is worth noting that, at this stage, the juvenile hormone has ceased to control the growth of the solution and the intensification process is performed till a maximum number of iterations are reached or when there is no significant change in the fitness of the current solution.

2.4.3.2 Postprocessing

The postprocessing operator is largely user guided; it allows the decision maker to interactively make expert changes to the current solution, and to rerun the intensification operator. As such, the termination of the maturation phase is determined by the user. This also ensures that expert knowledge and intuition are incorporated into the solution procedure, which significantly enhances the interactive search power of the algorithm.

2.5 FSM AND RELATED ALGORITHMS

FSM has a number of advantages over other related algorithms. Contrary to simulated annealing (SA), which makes purely random choices to decide the next move, FSM employs an intelligent selection operation to decide which changes to perform. Furthermore, FSM takes advantage of multiple transformation operations on weak elements of the current solution, allowing for more distant changes between successive iterations.

FSM, like the genetic algorithm (GA), uses the mechanics of evolution as it progresses from one generation to the other. The GA necessarily keeps a number of candidate solutions in each generation as parents, generating offspring by a crossover operator. On the other hand, FSM simulates metamorphosis, evolving a single solution under hormonal control. In addition, domain-specific heuristics are employed to regenerate and repair the emerging candidate solution, developing it into an improved and complete solution. In retrospect, FSM reduces the computation time needed to maintain a large population of candidate solutions in the GA.

The selection process in FSM is quite different from that in the GA and other related evolutionary algorithms. While the GA uses probabilistic selection to retain a set of good solutions from a population of candidate solutions, FSM selects and discards inferior elements of a candidate solution, according to the fitness of each element. This enhances the computational speed of the FSM procedure.

At the end of the growth phase, the FSM algorithm goes through a maturation phase where an intensive search process is performed to refine the solution, and possibly obtain an improved final solution. The algorithm allows the decision maker to input his or her managerial choices to guide the search process. This interactive facility gives FSM an added advantage over other heuristics.

The proposed algorithm uses hormonal control to enhance and guide its global multiobjective optimization process. This significantly eliminates unnecessary searches through regions with inferior solutions, hence improving the search efficiency of the algorithm. In summary, the previously mentioned advantages provide the FSM algorithm enhanced convergence characteristics that enable the algorithm to perform fewer computations relative to other evolutionary algorithms.

2.6 APPLICATION AREAS

The FSM approach is a multicriteria optimization algorithm that can be used as a decision support tool in several areas (see Table 2.1). Some of the areas are vehicle routing problems (Tarantilis et al., 2004), job shop scheduling (Sun et al., 1995; Sakawa and Kubota, 2000; Cheng et al., 2007), driver scheduling (Li and Kwan, 2001), crew scheduling (Ernst et al., 2001, Dawid, König, and Strauss, 2001), and healthcare staff scheduling (Jan et al., 2000; Cheang et al., 2003; Ernst et al., 2004; Mutingi and Mbohwa, 2014).

TABLE 2.1
Typical FSM Application Areas and the Associated Fuzzy Multiple Criteria Decisions

No.	Problem Area	Brief Description	Fuzzy Decision Criteria	Selected References
1	Vehicle routing	Serving clients at desired time windows with a fleet of vehicles	Satisfy time window preferences, client preferences, workload balance or fairness, and management desires and goals. Minimize costs, distance traveled, and time window violations.	Teodorovic and Pavkovic (1996); Tarantilis et al. (2004); Tang et al. (2009)
2	Job shop scheduling	Assigning a set of jobs on a set of machines where each job has a specified processing order through all machines	Satisfy due-date specifications, time window preferences, customer expectations, and management goals and choices. Minimize makespan, idle time, and production costs.	Subramaniam et al. (2000); Chen, Luh, and Fang (2001); Aydin and Fogarty (2004); Sakawa and Kubota (2000)
3	Driver scheduling	Allocation of shifts to drivers so as to cover all the work or trips required	Satisfy time windows, workload balance, driver preferences, and management choices. Minimize costs, distance traveled.	Li and Kwan (2001, 2003); Kwan, Kwan, and Wren (1999)
4	Crew scheduling	Sequencing and assignment of tasks (e.g., flight legs) to crew members so that all tasks are covered	Satisfy crew preferences, passenger satisfaction, management goals, and choices. Minimize costs, crew overcoverage and undercoverage.	Teodorovic (1998); Dawid et al. (2001); Ernst et al. (2004)
5	Healthcare staff scheduling	Construction and assignment of work schedules to healthcare workers	Satisfy shift preferences, workload balance or fairness, management choices, patient expectations, time window. Minimize costs, overstaffing and understaffing.	Cheang et al. (2003); Ernst et al. (2004); Topaloglu and Selim (2010); Rasmussen et al. (2012)

Vehicle routing. This problem is concerned with finding efficient routes with minimum total cost for a given fleet of vehicles that serve some commodities to a given number of customers. The problem can be more complicated in the presence of fuzzy time windows (Tang et al., 2009), fuzzy travel times, and fuzzy demands (Erbao and Mingyong, 2010; Marinakis, Iordanidou, and Marinaki, 2013). Analogous to the vehicle routing problem with time windows is the homecare nurse scheduling problem. For a more realistic decision approach, a multicriteria decision method such as FSM is promising.

Job shop scheduling. The process of dispatching jobs is extremely complex, especially when dynamic uncertainties exist in job arrival times or resource availability (Subramaniam et al., 2000). In some cases, the due dates of jobs are considered as time windows rather than as a point in time (Huang and Yang, 2008). With fuzzy time windows, the job shop scheduling problem becomes even more complex, especially when multiple criteria are taken into consideration. An analogous problem occurs in a hospital setting, where care tasks are assigned to a given number of nursing staff. FSM is a potential algorithm in these problem instances.

Driver scheduling. This is a process of partitioning blocks of work, each of which is serviced by one vehicle, into a set of legal driver shifts (Li and Kwan, 2001). The main objective is to minimize the total shift costs. Driver scheduling is a complex problem that is commonly solved using a set-covering approach (Li and Kwan, 2003). A large set of possible legal shifts is usually generated first, based on parameterized heuristics reflecting driver work rules and goals of individual companies. Personalized preferences and management goals need to be considered when constructing the schedules. Fuzzy multicriteria decision making using FSM could be more appropriate.

Crew scheduling. Crew scheduling and rostering are aimed at constructing and sequencing feasible pairings or duties from given timetables so that all trips are adequately covered (Ernst et al., 2004). Personalized rosters are then assigned to individual crews, while satisfying individual preferences and management goals. The ensuing problem is often overconstrained with imprecise multiple objectives. As such, a fuzzy multicriteria approach is expected to produce more realistic results efficiently.

Healthcare staff scheduling. The main research focus in healthcare staff scheduling has been in nurse scheduling. Other specific areas are homecare staff scheduling and physician scheduling. These healthcare staff scheduling problems are associated with multiple criteria targeted at providing adequate and suitably qualified healthcare staff to cover the demand arising from patients while satisfying employee preferences, management goals, work regulations, and fair distribution of night shifts, weekend shifts, and days off. In most cases, these multiple criteria are imprecise and the resulting scheduling problems are highly constrained (Cheang et al., 2003; Burke et al., 2004). FSM may come in handy for such problem instances.

Other areas. Other application areas include digital circuit design (Kling and Banejee, 1987; Saiti, Youssef, and Ali, 1999; Saiti and Al-Ismail, 2004), production planning problems (Vasant and Barsoum, 2009, 2010), and other operations management problems in call centers, civic services, emergency services, retail systems, and construction (Vasant, 2010).

All these problem areas are associated with multiple conflicting objectives, imprecise fuzzy goals and constraints, and the need for interactive optimization approaches that can incorporate the choices, intuitions, and expert judgments of the decision maker. Of particular interest, staff scheduling, predominantly nurse scheduling, has attracted the attention of most researchers in the operations research and operations management community.

2.7 SUMMARY

This research presented a fuzzy simulated metamorphosis algorithm, motivated by the need for efficient fuzzy multicriteria decision support and the biological metamorphosis evolution common in insects. The algorithm mimics the hormone-controlled evolution process going through initialization, an iterative growth loop, and, finally, a maturation loop.

The proposed method is a realistic approach to optimizing multicriteria decision problems with fuzzy conflicting goals and constraints such as nurse scheduling, homecare nurse scheduling, vehicle routing, care task assignment, and job shop scheduling. Equipped with fuzzy evaluation techniques and the evolutionary metamorphosis concepts, the algorithm offers an interactive approach that can incorporate the decision maker's expert choices, intuition, and experience. The algorithm works on a single candidate solution to efficiently search for the best solution using two iterative loops with unique operators, under hormonal guidance.

REFERENCES

Aydin, M. E. and Fogarty, T. C. 2004. A simulated annealing algorithm for multi-agent systems: A job shop scheduling application. *Journal of Intelligent Manufacturing* 15: 805–814.

Burke, E., De Causmaecker, P., Vander Berghe, G. and Landeghem H. 2004. The state of the art of nurse rostering. *Journal of Scheduling* 7: 441–499.

Cheang, B., Li, H., Lim, A. and Rodrigues, B. 2003. Nurse rostering problems—A bibliographic survey. *European Journal of Operational Research* 151: 447–460.

Chen, H., Luh, P. B. and Fang, L. 2001. A time window based approach for job shop scheduling. *IEEE International Conference on Robotics & Automation*, Seoul, Korea, 842–847.

Cheng, M., Ozaku, H. I., Kuwahara, N., Kogure, K. and Ota, J. 2007. Nursing care scheduling problem: Analysis of staffing levels. *IEEE Proceedings of the 2007 International Conference on Robotics and Biomimetics* 1: 1715–1719.

Dawid, H., Konig, J. and Strauss, C. 2001. An enhanced rostering model for airline crews. *Computers and Operations Research* 28: 671–688.

Erbao, C. and Mingyong, L. 2010. The open vehicle routing problem with fuzzy demands. *Expert Systems with Applications* 37: 2405–2411.

Ernst, A., Jiang, H., Krishnamoorthy, M., Nott, H. and Sier, D. 2001. An integrated optimization model for train crew management. *Annals of Operations Research* 108: 211–224.

Ernst, A. T., Jiang, H., Krishnamoorthy, M. and Sier, D. 2004. Staff scheduling and rostering: A review of applications, methods and models. *European Journal of Operational Research* 153: 3–27.

Huang, R.-H. and Yang C.-L. 2008. Ant colony system for job shop scheduling with time windows. *International Journal of Advanced Manufacturing Technology* 39: 151–157.

Inoue T., Furuhashi T., Maeda H. and Takaba, M. 2003. A proposal of combined method of evolutionary algorithm and heuristics for nurse scheduling support system. *IEEE Transactions on Industrial Electronics* 50 (5): 833–838.

Jan, A., Yamamoto, M. and Ohuchi, A. 2000. Evolutionary algorithms for nurse scheduling problem. *IEEE Proceedings of the 2000 Congress on Evolutionary Computation* 1: 196–203.

Kling, R. M. and Banejee, P. 1987. ESP: A new standard cell placement package using simulated evolution. *Proceedings of the 24th ACWIEEE Design Automation Conference*, 60–66.

Kwan, A. S. K., Kwan, R. S. K. and Wren, A. 1999. Driver scheduling using genetic algorithms with embedded combinatorial traits, in *Computer-aided transit scheduling*, ed. N. H. M. Wilson, 81–102. Springer–Verlag, Berlin.

Li, J. and Kwan, R. S. K. 2001. A fuzzy simulated evolution algorithm for the driver scheduling problem. *Proceedings of the 2001 IEEE Congress on Evolutionary Computation, IEEE Service Center*, 1115–1122.

Li, J. and Kwan, R. S. K. 2003. A fuzzy genetic algorithm for driver scheduling. *European Journal of Operational Research* 147: 334–344.

Marinakis, Y., Iordanidou, G-R. and Marinaki, M. 2013. Particle swarm optimization for the vehicle routing problem with stochastic demands. *Applied Soft Computing* 13: 1693–1704.

Mutingi, M. and Mbohwa, C. 2014. Healthcare staff scheduling in a fuzzy environment: A fuzzy genetic algorithm approach. *Proceedings of the 2014 International Conference on Industrial Engineering and Operations Management*, Bali, Indonesia, 303–312.

Osman, I.H. and Laporte, G. 1996. Metaheuristics: A bibliography. *Annals Operations Research* 63: 513–623.

Rasmussen, M. S., Justesen, T., Dohn, A. and Larsen, J. 2012. The home care crew scheduling problem: Preference-based visit clustering and temporal dependencies. *European Journal of Operational Research* 219: 598–610.

Saiti, S. M. and Al-Ismail, M. S. 2004. Enhanced simulated evolution algorithm for digital circuit design yielding faster execution in a larger solution space. *IEEE Congress on Evolutionary Computation, CEC2004* 2: 1794–1799.

Saiti, S. M., Youssef, H. and Ali, H. 1999. Fuzzy simulated evolution algorithm for multi-objective optimization of VLSI placement. *Congress on Evolutionary Computation*, 91–97.

Sakawa, M. and Kubota, R. 2000. Fuzzy programming for multi-objective job shop scheduling with fuzzy processing time and fuzzy due date through genetic algorithms. *European Journal of Operational Research* 120 (2): 393–407.

Shaffer, S. 1991. A rule-based expert system for automated staff scheduling. *IEEE International Conference on Systems, Man, and Cybernetics* 3: 1691–1696.

Subramaniam, V., Ramesh, T., Lee, G. K., Wong, Y. S. and Hong, G. S. 2000. Job shop scheduling with dynamic fuzzy selection of dispatching rules. *International Journal of Advanced Manufacturing Technology* 16: 759–764.

Sun, D., Batta, R. and Lin, L. 1995. Effective job shop scheduling through active chain manipulation. *Computers & Operations Research* 22 (2): 159–172.

Tang, J., Pan, Z., Fung, R. Y. K. and Lau, H. 2009. Vehicle routing problem with fuzzy time windows. *Fuzzy Sets and Systems* 160: 683–695.

Tarantilis, C. D., Kiranoudis, C. T. and Vassiliadis, V. S. A. 2004. A threshold accepting metaheuristic for the heterogeneous fixed fleet vehicle routing problem. *European Journal of Operations Research* 152: 148–158.

Teodorovic, D. 1998. A fuzzy set theory approach to the aircrew rostering. *Fuzzy Sets and Systems* 95 (3): 261–271.

Teodorovic, D. and Pavkovic, G. 1996. The fuzzy set theory approach to the vehicle routing problem when demand at nodes is uncertain. *Fuzzy Sets and Systems* 82 (3): 307–317.

Topaloglu, S. and Selim, S. 2010. Nurse scheduling using fuzzy modeling approach. *Fuzzy Sets and Systems* 161: 1543–1563.

Truman, J. W. and Riddiford, L. M. 2002. Endocrine insights into the evolution of metamorphosis in insects. *Annual Review of Entomology* 47: 467–500.

Tufte, G. 2011. Metamorphosis and artificial development: An abstract approach to functionality. In *ECAL 2009, Part I, LNCS* 5777, eds. G. Kampis, I. Karsai, and E. Szathmary, 83–90, Springer–Verlag, Berlin.

Vasant, P. 2010. Hybrid simulated annealing and genetic algorithms for industrial production management problems. *International Journal of Computational Methods* 7 (2): 279–297.

Vasant, P. and Barsoum, N. 2009. Hybrid genetic algorithms and line search method for industrial production planning with non-linear fitness function. *Engineering Applications of Artificial Intelligence* 22 (4–5): 767–777.

Vasant, P. and Barsoum, N. 2010. Hybrid pattern search and simulated annealing for fuzzy production planning problems. *Computers and Mathematics with Applications* 60 (4): 1058–1067.

3 Fuzzy Simulated Evolution Algorithm

3.1 INTRODUCTION

Modern metaheuristic approaches, particularly biologically inspired evolutionary algorithms, have attracted the attention of many researchers concerned with multi-criteria decision making in various disciplines (Senvar, Turanoglu, and Kahraman, 2013). Some of the most popular algorithms are genetic algorithms, neural networks, particle swarm intelligence, the ant colony algorithm, and the simulated evolution algorithm. Significant research activities have implemented these algorithms with appreciable results (Senvar, Turanoglu, and Kahraman, 2013). However, when addressing complex multicriteria decision problems under fuzziness, fuzzy evaluation techniques are an essential addition, if more realism is desired in the algorithm chosen.

Fuzzy evaluation techniques accommodate imprecision, uncertainty, or partial truth, based on fuzzy theory concepts (Zimmerman, 1993; Vasant, 2013). In addition, these techniques can also handle real-world problems with multiple criteria. An important research direction is hybridizing efficient evolutionary approaches with fuzzy evaluation concepts (Mutingi and Mbohwa, 2014). The goal is to develop hybrid fuzzy evolutionary algorithms to provide optimal or near-optimal solutions within a reasonable computation time.

In this chapter, a fuzzy multicriteria evaluation approach is developed based on fuzzy set theory concepts. The approach is hybridized with the simulated evolution algorithm to come up with a fuzzy simulated evolution algorithm. In this regard, the purpose of this chapter is to present a fuzzy simulated evolution algorithm for solving complex multicriteria decision problems under fuzziness. Hence, the specific research objectives are as follows:

1. To present a background to the basic simulated evolution algorithm
2. To present a fuzzy multicriteria evaluation approach based on fuzzy theory concepts
3. To develop a fuzzy simulated evolution algorithm framework

The rest of the chapter is structured as follows. The next section briefly describes the basic simulated evolution algorithm. This is followed by an outline of fuzzy multicriteria evaluation. A fuzzy simulated evolution algorithm is then presented, followed by an outline of potential application areas.

3.2 SIMULATED EVOLUTION ALGORITHM

Simulated evolution (SE) is an evolutionary optimization approach originally proposed by Kling and Banerjee (1987). Inspired by the philosophy of natural selection in biological environments, the SE algorithm evolves a single candidate solution from one generation (iteration) to the next by eliminating or discarding inferior elements in the solution. Thus, in each generation, elements with high fitness under current conditions are retained. The desired goal is to gradually create a stable solution perfectly adapted to the given constraints. To escape from local optima, mutation perturbs genetic inheritance in anticipation of new, improved genetic information, enabling the algorithm to effectively explore and exploit the solution space (Shiraishi, Ono, and Dahb, 2009).

A few applications of the SE algorithm exist in the literature. Ly and Mowchenko (1993) developed a simulated evolution approach that can effectively explore and exploit the solution space for high-level synthesis. Sait and AI-Ismail (2004) developed an enhanced simulated evolution algorithm for digital circuit design for faster execution in a larger solution space. In the same vein, Shiraishi, Ono, and Dahb (2009) presented a solution space reduction procedure of a simulated evolution algorithm for solving standard cell placement problem.

The SE procedure consists of *evaluation*, *selection*, and *reconstruction* operators that iteratively work on a single candidate solution. Prior to evaluation, initialization creates a valid starting solution and accepts input parameters. The evaluation operator then computes the fitness of each element in the solution, which is used to probabilistically select and discard weak elements. The resulting incomplete solution is rebuilt by the reconstruction operator using problem-specific heuristics. The complete solution is then passed on to the evaluation operator, repeating the procedure until a termination condition is fulfilled. A summary of the SE procedure is presented in Figure 3.1.

The basic SE procedure is a search and optimization heuristic that improves the solution through iterative perturbation and reconstruction. However, the iterative process ensures that the best solution is always preserved. To enhance its search

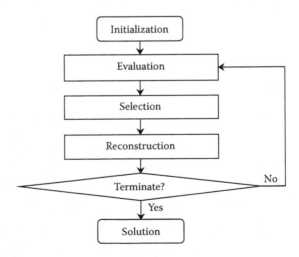

FIGURE 3.1   A flowchart of the basic SE procedure.

and optimization efficiency in a fuzzy multicriteria environment, SE needs to incorporate fuzzy evaluation techniques. The next section presents the fuzzy simulated evolution approach for this purpose.

3.3 FUZZY SIMULATED EVOLUTION ALGORITHM

The fuzzy simulated evolution (FSE) algorithm is developed from the basic SE algorithm. Several stages of the algorithm can be fuzzified—for instance, initialization, evaluation, and selection. A few applications of FSE are found in the literature. Sait and AI-Ismail (2004) used a fuzzy simulated evolution algorithm for finding near-optimal solutions to a large-scale multiobjective optimization placement problem. Li and Kwan (2001) developed a fuzzy simulated evolution algorithm for the driver scheduling problem. For ease of deliberation, key definitions and terminologies need to be clarified.

> **Definition 1:** *Element.* A solution of an optimization problem, represented by a suitable code, can be split into subunits called elements. Each element j ($j = 1,2,\ldots,n$) represents a solution to a subproblem of the optimization problem.
>
> **Definition 2:** *Candidate solution.* A candidate solution, s_t, is defined as the most suitable combination of the elements at iteration t. This implies that $s_t \subseteq E$, where E is the solution space of the problem.

As an example, consider a nurse rostering problem consisting of a group of nurses to be scheduled over a period of 1 week. Each nurse is allocated a single shift schedule spanning a week. Specific allowable shifts are (1) day shift d, covering period 0800–1600 hours, (2) night shift n covering period 1600–2400 hours, (3) late night shift l covering period 0000–0800 hours, and (4) day off o.

Figure 3.2 shows a typical roster for a nurse rostering problem. In this case, the complete roster is a candidate solution, while single shift schedules represent elements.

Nurse	Skill	Days 1	2	3	4	5	6	7	d	n	l
s_1	1	l	n	l	n	d	l	n	1	3	3
s_2	1	o	d	n	l	d	d	l	3	1	2
s_3	1	d	d	d	d	o	n	d	5	1	0
s_4	2	n	l	l	o	d	d	d	3	1	2
s_5	2	d	d	h	n	n	l	n	2	1	1
s_6	2	d	o	d	d	l	n	d	4	1	1
s_7	2	l	n	d	d	l	d	o	3	1	2
s_8	2	n	l	n	l	n	o	l	0	3	3
d		3	3	3	3	3	3	3			
n		2	2	2	2	2	2	2			
l		2	2	2	2	2	2	2			

FIGURE 3.2 A typical candidate solution for the nurse rostering problem.

3.3.1 INITIALIZATION

In initialization, a good initial solution is generated as a seed for succeeding iterations. The fitness of the seed may influence the overall computation time of the algorithm. The following approaches can be used to generate good initial solutions:

1. Random generation of the initial solution
2. A greedy approach using a guided probabilistic bias
3. Domain-specific constructive heuristics guided by problem constraints
4. User-defined candidate solution

Following the initialization phase, the FSE iteratively loops through evaluation, selection, and reconstruction heuristics until a termination criterion is satisfied. Possible termination criteria can be implemented based on (1) a predefined number of iterations, as specified by the decision maker, or (2) a preset limit to the number of iterations performed without significant solution improvement.

3.3.2 EVALUATION

Fuzzy evaluation determines the goodness or fitness of the structure of each element in the solution. The aim is to provide an evaluation of the contribution of each element to the overall fitness of the candidate solution so that elements contributing much less than is acceptable can be identified.

The fitness λ_j of each element j in the current solution s_t at iteration t is evaluated based on a normalized function, which may be a combination of normalized membership functions. Thus, in the case where an element has to be evaluated using n membership functions, $\mu_1, \mu_2, ..., \mu_n$, the overall fitness for each element can be formulated using a weighted multicriteria approach:

$$\lambda_j = \sum_{i=1}^{n} w_i \mu_i \tag{3.1}$$

where w_i is the weight of each objective function i, such that $\Sigma w_i = 1.0$. These weights enable decision makers to incorporate their choices to reflect fuzzy human preferences and intuitions. The overall fitness F_t of the candidate solution at iteration t is calculated in terms of individual fitness of all the elements as follows:

$$F_t = \lambda_1 \wedge \lambda_2 \wedge ... \wedge \lambda_n \tag{3.2}$$

where the symbol "$\wedge$" represents the fuzzy min operator (Bellman and Zadeh, 1970; Klir and Yuan, 1995).

3.3.3 SELECTION

Selection probabilistically determines whether an element j should be retained into the next generation; an element j with a high fitness value has a higher probability

Selection algorithm		
1.	Set constant p;	
2.	**Begin**	
3.	Step 1. Compute fitness λ_j of element j;	
4.	Step 2. Let p_t = Random $[0,1]$;	
5.	Step 3. Let λ_j = max $[0, p_t - p]$;	
6.	Step 4. **If** $(\lambda_j < \lambda_t)$ **Then** discard j, **Return**	
7.	**Else** return j	
8.	**End If**	
9.	**End**	

FIGURE 3.3 An algorithm for the selection phase.

of surviving into the next generation. Discarded elements are reserved in a set Q for the reconstruction phase. Selection compares fitness λ_j with an allowable fitness λ_t at iteration t:

$$\lambda_t = \max [0, p_t - p] \tag{3.3}$$

where p_t is a random number in $[0,1]$ at iteration t and p is a predetermined constant in $[0,1]$.

Figure 3.3 summarizes the selection algorithm. The algorithm computes the allowable fitness λ_t. At each iteration t, compare λ_j with λ_t, and return the element with better fitness.

The expression $\lambda_t = p_t - p$ facilitates convergence. Consequently, the search ability of the algorithm can be controlled by setting the value of p to a suitable value.

3.3.4 MUTATION

Mutation enhances intensive search in the neighborhood of the current solution while allowing for explorative search in unvisited regions of the solution space. Intensification randomly swaps chosen pairs of elements. On the other hand, exploration enables the algorithm to move from local optima by probabilistically eliminating some elements, even the best performing ones. Generally, mutation is applied at a very low probability p_m to ensure convergence. A decay function may be used for dynamic mutation as follows:

$$p_m(t) = p_0 e^{-\varepsilon t/T} \tag{3.4}$$

where
   t is the iteration count
   T is the maximum count
   p_0 is the initial mutation probability
   ε is an adjustment factor

This expression can be used for both explorative and intensive mutation probabilities. Any infeasible partial solutions are repaired in the reconstruction phase.

Overall FSE algorithm
1. **Begin**
2.   Input; p, p_m, p_0, T;
3.   Initialize; randomly generate solution S;
4.   **Repeat**
5.   Evaluation;
6.   Selection;
7.   Mutation;
8.   Reconstruction;
9.   Termination condition, $t = t + 1$;
10.   **Until** (Termination criteria are satisfied, $t \geq T$)
11.   **Return** best solution s^*;
12. **End**

FIGURE 3.4   The FSE overall algorithm.

3.3.5   RECONSTRUCTION

The aim of reconstruction is to rebuild the partial solution from the previous stages into a complete solution. Values are added to empty spaces in the incomplete solution, using a greed-based heuristic, assuming that the attractiveness of adding an element j increases fitness λ_j in that solution. Discarded elements from previous generations, preserved in set Q, may be used for reconstruction.

3.3.6   THE OVERALL FSE ALGORITHM

The overall structure of the FSE pseudo-code is made up of the algorithms described in the previous sections. The algorithm obtains input from the user: p, p_m, p_0, and T. Following the initialization algorithm, the algorithm loops through evaluation, selection, mutation, and reconstruction. Figure 3.4 presents the overall structure of the FSE algorithm.

3.4   FSE AND RELATED ALGORITHMS: A COMPARISON

The FSE algorithm has a number of advantages over related metaheuristic approaches, such as genetic algorithm (Man, Tang, and Kwong, 1999; Sakawa and Kato, 2003), particle swarm optimization (Kennedy and Eberhart, 1995; Lo and Lin, 2011), and simulated annealing (Kirkpatrick, Gelatt, and Vecchi, 1983). Some of the advantages realized in this chapter are outlined as follows:

1. The algorithm mimics evolutionary principles based on a single candidate solution, which avoids excessive computation time.
2. The procedure of the algorithm is easy to construct and implement; it can be adapted to specific problem areas without much difficulty.
3. Unlike genetic algorithms, the algorithm selects inferior elements of only one solution to discard, in accordance with the fitness of each element.

4. The algorithm has stronger convergence capabilities, leading to less computation time than competitive metaheuristics algorithms such as genetic algorithms.

5. The algorithm incorporates features of probabilistic hill climbing, which gives the algorithm the power to explore unvisited regions of the solution space.

In consideration of the preceding advantages, it can be argued that the FSE algorithm is a potentially effective and efficient solution approach to several problem areas.

3.5 APPLICATION AREAS

FSE is a multicriteria approach that can be used for decision support in several areas—for instance, nurse scheduling (Jan, Yamamoto, and Ohuchi, 2000; Cheang et al., 2003; Ernst et al., 2004), driver scheduling (Li and Kwan, 2001, 2003), vehicle routing problems (Tarantilis, Kiranoudis, and Vassiliadis, 2004), and crew scheduling (Dawid, Konig, and Strauss, 2001). Major areas and their respective descriptions are listed in Table 3.1.

Nurse scheduling. Nurse rostering or scheduling is concerned with construction and allocation of shift schedules to a given set of nurses. The problem is often associated with multiple criteria, including adequate demand coverage, patient satisfaction, staff satisfaction, meeting management goals and work regulations, and fair distribution of shifts. Oftentimes, these multiple criteria are fuzzy and highly constrained (Cheang et al., 2003; Ernst et al., 2004).

Vehicle routing. The aim of solving the vehicle routing problem is to find efficient routes with minimum cost for a fleet of vehicles that serve a given number of clients. The problem is often inundated with fuzzy parameters such as time windows (Tang et al., 2009), travel times, and demands (Erbao and Mingyong, 2010; Brandao, 2011). The multicriteria FSE approach offers a more promising and realistic decision method than conventional methods.

Job shop scheduling. Job dispatching in an uncertain environment with imprecise arrival times and resource availability can be complex (Subramaniam et al., 2000). In some cases, the due dates of jobs are expressed as time windows (Huang and Yang, 2008). In the presence of fuzzy time windows, the problem becomes even more complex, considering its multicriteria nature. FSE is a potential algorithm in such problem situations.

Driver scheduling. This problem entails partitioning blocks of work, each of which is served by one vehicle, into a set of legal driver shifts (Li and Kwan, 2003). The aim is to minimize the total shift costs. The problem can be solved as a set covering problem, where a large set of possible legal shifts is generated based on parameterized heuristics reflecting driver work rules and goals of individual companies (Li and Kwan, 2001). However, personalized preferences and management goals need to be considered when constructing the schedules. Therefore, fuzzy multicriteria decision-making methods, such as FSE, can be more appropriate.

TABLE 3.1

Potential FSE Application Areas and Associated Fuzzy Decision Criteria

No.	Problem Area	Brief Description	Fuzzy Decision Criteria	Selected References
1	Nurse scheduling	Construction and assignment of work schedules to healthcare workers	To satisfy nurse preferences, workload balance, management choices To satisfy patient expectations, time windows To minimize costs, overstaffing, and understaffing	Cheang et al. (2003); Topaloglu and Selim (2010); Mutingi and Mbohwa (2013)
2	Driver scheduling	Allocation of shifts to drivers so as to cover all the required trips	To satisfy driver preferences, workload balance To satisfy client time window preferences To satisfy management choices, minimize costs, and distance traveled	Kwan, Kwan, and Wren (1999); Li and Kwan (2001, 2003)
3	Vehicle routing	Serving clients at their desired time windows with a given fleet of vehicles	To satisfy client preferences, workload balance To satisfy client time window preferences To satisfy management desires and goals, minimize costs and distance traveled	Teodorovic and Pavkovic (1996); Tarantilis, Kiranoudis, and Vassiliadis (2003, 2004); Tang et al. (2009)
4	Job shop scheduling	Assigning a set of jobs on a set of machines where each job has a specified processing order through all machines	To satisfy due-date specifications, time window preferences, customer expectations, and management goals and choices To minimize makespan, idle time, and production costs	Sakawa and Kubota (2000); Subramaniam et al. (2000); Chen, Luh, and Fang (2001)
5	Crew scheduling	Sequencing and assigning tasks (e.g., flight legs) to crew members to cover all tasks are covered	To satisfy crew preferences To satisfy passenger expectations To satisfy management goals and choices, minimize costs, and minimize crew over-/undercoverage	Teodorovic (1998); Dawid, Konig, and Strauss (2001); Ernst et al. (2004)

Crew scheduling. The aim of crew scheduling and rostering is to construct a set of feasible pairings from given timetables so that all trips are adequately covered (Ernst et al., 2004). Personalized rosters are then assigned to individual crews. Most of the problem instances are concerned with satisfying imprecise individual crew preferences, passenger expectations, and management goals. As such, a fuzzy multicriteria approach such as FSE comes in handy.

Other areas. Other possible areas of application are physician scheduling (Lo and Lin, 2011), homecare worker scheduling (Mutingi and Mbohwa, 2013), production planning and control (Vasant, 2006, 2011), digital circuit design (Kling and Banejee, 1987; Sait, Youssef, and Ali, 1999; Sait and Al-Ismail, 2004), and operations management (Vasant, 2013).

All these mentioned problems are multicriteria decision problems with fuzzy goals and constraints. They demand solution approaches that can accommodate the choices and intuitions of the decision maker. For instance, nurse scheduling in a fuzzy environment is a common challenge in most healthcare settings.

3.6 SUMMARY

Developing efficient algorithms for solving complex multicriteria decision problems is imperative. Decision makers continue to face more complex problems in various sectors of industry, such as healthcare, manufacturing, transportation, and logistics. This chapter presented a fuzzy simulated evolution algorithm. The algorithm incorporates the concepts of evolution, fuzzy set theory, iterative improvement, and constructive perturbation into the classical simulated evolution algorithm.

A number of advantages can be observed from the FSEA algorithm presented in this chapter. The algorithm is easy to construct and implement, it works on a single candidate solution rather than on a population of solutions, it has strong convergence ability, and it incorporates features of probabilistic hill climbing, giving the algorithm the power to explore unvisited regions of the solution space. To this effect, one interesting and potential area of application for this algorithm is nurse rostering and nurse rerostering.

REFERENCES

Bellman, R. E. and Zadeh, L. A. 1970. Decision making in a fuzzy environment. *Management Science* 17: 141–164.

Brandao, J. 2011. A tabu search algorithm for the heterogeneous fixed fleet vehicle routing problem. *Computers & Operations Research* 38 (1): 140–151.

Cheang, B., Li, H., Lim, A. and Rodrigues, B. 2003. Nurse rostering problems—A bibliographic survey. *European Journal of Operational Research* 151: 447–460.

Chen, H., Luh, P. B. and Fang, L. 2001. A time window based approach for job shop scheduling. *IEEE International Conference on Robotics & Automation*, Seoul, Korea, 842–847.

Dawid, H., Konig, J. and Strauss, C. 2001. An enhanced rostering model for airline crews. *Computers & Operations Research* 28. 671–688.

Erbao, C. and Mingyong, L. 2010. The open vehicle routing problem with fuzzy demands. *Expert Systems with Applications* 37: 2405–2411.

Ernst, A. T., Jiang, H., Krishnamoorthy, M. and Sier, D. 2004. Staff scheduling and rostering: A review of applications, methods and models. *European Journal of Operational
Research* 153: 3–27.

Huang, R.-H. and Yang, C.-L. 2008. Ant colony system for job shop scheduling with time
windows. *International Journal of Advanced Manufacturing Technology* 39: 151–157.

Jan, A., Yamamoto, M. and Ohuchi, A. 2000. Evolutionary algorithms for nurse scheduling
problem. *IEEE Proceedings of the 2000 Congress on Evolutionary Computation* 1:
196–203.

Kennedy, J. and Eberhart, R. C. 1995. Particle swarm optimization. *IEEE International
Conference on Neural Networks* 4: 1942–1948.

Kirkpatrick, S., Gelatt, C. D. and Vecchi, M. P. 1983. Optimization by simulated annealing.
Science, New Series 220: 671–680.

Kling, R. M. and Banejee, P. 1987. ESP: A new standard cell placement package using simulated evolution. *Proceedings of the 24th ACWIEEE Design Automation Conference*,
60–66.

Klir, G. J. and Yuan, B. 1995. *Fuzzy sets and fuzzy logic: Theory and applications*. Prentice
Hall, Upper Saddle River, NJ.

Kwan, A. S. K., Kwan, R. S. K. and Wren, A. 1999. Driver scheduling using genetic algorithms with embedded combinatorial traits, in *Computer-aided transit scheduling*,
ed. N. H. M. Wilson, 81–102. Springer-Verlag, Berlin.

Li, J. and Kwan, R. S. K. 2001. A fuzzy simulated evolution algorithm for the driver scheduling problem. *IEEE Congress on Evolutionary Computation*, 1115–1122.

Li, J. and Kwan, R. S. K. 2003. A fuzzy genetic algorithm for driver scheduling. *European Journal
of Operational Research* 147: 334–344.

Lo, C.-C. and Lin, T.-H. 2011. A particle swarm optimization approach for physician scheduling in a hospital emergency department. *IEEE Seventh International Conference on
Natural Computation*, 1929–1933.

Ly, T. A. and Mowchenko, J. T. 1993. Applying simulated evolution to high level synthesis.
IEEE Transactions on Computer-Aided Design of Integrated Circuits and Systems 12:
389–409.

Man, K. F., Tang, K. S. and Kwong, S. 1999. *Genetic algorithms: Concepts and design*,
Springer, London.

Mutingi, M. and Mbohwa, C. 2013. A fuzzy simulated evolution algorithm for multi-objective
homecare worker scheduling. *IEEE International Conference on Industrial Engineering
and Engineering Management*, Thailand, 10–13 December 2013.

Mutingi, M. and Mbohwa, C. 2014. Healthcare staff scheduling in a fuzzy environment: A
fuzzy genetic algorithm approach. *Proceedings of the 2014 International Conference on
Industrial Engineering and Operations Management*, Bali, Indonesia, 303–312.

Sait, S. M. and AI-Ismail, M. S. 2004. Enhanced simulated evolution algorithm for digital
circuit design yielding faster execution in a larger solution space. *IEEE Congress on
Evolutionary Computation, CEC2004* 2: 1794–1799.

Sait, S. M., Youssef, H. and Ali, H. 1999. Fuzzy simulated evolution algorithm for multi-objective
optimization of VLSI placement. *Congress on Evolutionary Computation*, 91–97.

Sakawa, M. and Kato, K. 2003. Genetic algorithms with double strings for 0-1 programming
problems. *European Journal of Operational Research* 144: 582–597.

Sakawa, M. and Kubota, R. 2000. Fuzzy programming for multi-objective job shop scheduling
with fuzzy processing time and fuzzy due date through genetic algorithms. *European
Journal of Operational Research* 120 (2): 393–407.

Senvar, O., Turanoglu, E. and Kahraman, C. 2013. Usage of metaheuristics in engineering: A
literature review. In *Meta-heuristics optimization algorithms in engineering, business,
economics, and finance*, ed. P. Vasant, Information Science Reference, Hershey, PA,
pp. 484–528.

Shiraishi, Y., Ono, T. and Dahb, M. A. E. 2009. Solution space reduction of simulated evolution algorithm for solving standard cell placement problem. *IEEE Fifth International Conference on Natural Computation*, 420–424.

Subramaniam, V., Ramesh, T., Lee, G. K., Wong, Y. S. and Hong, G. S. 2000. Job shop scheduling with dynamic fuzzy selection of dispatching rules. *International Journal of Advanced Manufacturing Technology* 16: 759–764.

Tang, J., Pan, Z., Fung, R. Y. K. and Lau, H. 2009. Vehicle routing problem with fuzzy time windows. *Fuzzy Sets and Systems* 160: 683–695.

Tarantilis, C. D., Kiranoudis, C. T. and Vassiliadis, V. S. A. 2003. A list based threshold accepting metaheuristic for the heterogeneous fixes fleet vehicle routing problem. *Journal of the Operational Research Society* 54 (1): 65–71.

Tarantilis, C. D., Kiranoudis, C. T. and Vassiliadis, V. S. A. 2004. A threshold accepting metaheuristic for the heterogeneous fixed fleet vehicle routing problem. *European Journal of Operations Research* 152: 148–158.

Teodorovic, D. 1998. A fuzzy set theory approach to the aircrew rostering. *Fuzzy Sets and Systems* 95 (3): 261–271.

Teodorovic, D. and Pavkovic, G. 1996. The fuzzy set theory approach to the vehicle routing problem when demand at nodes is uncertain. *Fuzzy Sets and Systems* 82 (3): 307–317.

Topaloglu, S. and Selim, S. 2010. Nurse scheduling using fuzzy modeling approach. *Fuzzy Sets and Systems* 161: 1543–1563.

Vasant, P. 2006. Fuzzy production planning and its application to decision making. *Journal of Intelligent Manufacturing* 17 (1): 5–12.

Vasant, P. 2011. Hybrid MADS and GA techniques for industrial production systems. *Archives of Control Sciences* 21 (3): 227–240.

Vasant, P. 2013. Hybrid LS-SA-PS methods for solving fuzzy non-linear programming problems. *Mathematical and Computer Modeling* 57 (1–2): 180–188.

Zimmerman, H. J. 1993. *Fuzzy set theory and its applications*, 2nd rev. ed. Kluwer Academic Publishers: Dordrecht.

4 Fuzzy Grouping Genetic Algorithm

4.1 INTRODUCTION

Service and manufacturing systems are inundated with problem situations where system entities are required to be grouped into clusters so as to achieve a certain goal (Onwubolu and Mutingi, 2001; Mutingi and Mbohwa, 2014). Typically, the concept of grouping the entities is aimed at optimizing system efficiency and effectiveness. For instance, when dispatching homecare nurses to serve patients at their homes, the decision maker has to allocate a group of patients to each nurse such that patient visits and the homecare service are done as efficiently and effectively as possible.

Analogous to the homecare nurse scheduling is the vehicle routing problem, a common problem in logistics and the transport industry (Mutingi and Mbohwa, 2012, 2013a). To minimize transportation costs, the number of vehicles used, and customer waiting times, the assignment of groups of clients to be visited should be optimized (Mutingi and Mbohwa, 2012). It is therefore crucial to know how to construct efficient groups of clients to be visited, considering the size, type, and capacity of the available vehicles (Taillard, 1999; Toth and Vigo, 2002; Tarantilis, Kiranoudis, and Vassiliadis, 2004). In addition, when assigning tasks to workers, such as nurses in a hospital setting, it is crucial to know how to create groups of tasks and assign them to workers as effectively as possible.

In a manufacturing setting, managers always desire to find the most effective way to group parts with similar characteristics (Onwubolu and Mutingi, 2001). The aim is to schedule similar parts to ensure that specific groups are produced using specific processes in specific departments.

The previously mentioned problems are of common occurrence in various types of industry, from manufacturing to the service industry. For the purpose of this research, these are called *grouping problems*. It is important to note that these problems are inherently difficult to solve because of their combinatorial nature (Onwubolu and Mutingi, 2001; Filho and Tiberti, 2006). A literature search survey on extant grouping problems revealed notable characteristics as outlined here:

1. The problems have a group structure that can be exploited in solution search and optimization.
2. The problems are highly combinatorial in nature, which makes them difficult to solve.
3. The problems are highly constrained, which adds to their complexity.
4. The problems have imprecise parameters involving human preferences and choices that are difficult to quantify.

In light of their complex characteristics, grouping problems are often solved using expert systems and heuristic and metaheuristic methods. Potential metaheuristic methods are genetic algorithms, particle swarm optimization, tabu search, and other evolutionary algorithms. These solution methods are designed to take advantage of the group structure of specific problems. This chapter focuses on the development of a group genetic algorithm approach for solving grouping problems. The objectives of the research are as follows:

1. To describe the basic grouping genetic algorithm as an extension of the classical genetic algorithm
2. To develop a fuzzy grouping genetic algorithm as an extension of the basic grouping genetic algorithm
3. To outline the potential application areas of the fuzzy grouping genetic algorithm

The next section explains the basic grouping genetic algorithm. This is followed by a description of the proposed fuzzy grouping genetic algorithm. Suggested application areas are then presented. The chapter ends with a summary and further research prospects.

4.2 GROUPING GENETIC ALGORITHM

The grouping genetic algorithm (GGA) is an extension of a genetic algorithm originally developed by Falkenauer (1992) to solve grouping problems. The algorithm consists of group chromosome representation, initial population generation, fitness evaluation, and an iterative loop comprising selection, crossover, and mutation operators equipped with enhancements to handle the group structure of the problems. Mutingi and Mbohwa (2012) developed an enhanced algorithm for the heterogeneous fixed fleet vehicle routing problem.

Figure 4.1 presents a flowchart for the basic GGA. The algorithm begins by generating an initial population of the desired size. The initialization process can be achieved through random generation or by using domain-specific greedy heuristics (Filho and Tiberti, 2006). For instance, customers, clients, or patients are randomly assigned to available nursing staff, while observing the capacities of each nurse. Specific heuristic procedures may be added to the initialization procedure in order to enhance the initial population generation. The population of chromosomes is then evaluated for fitness.

The GGA procedure computes the cost function $g(s)$ for each chromosome s and determines the chromosome with the maximum cost function value (Mutingi and Mbohwa, 2013a,b). The cost function is a combination of cost function values of individual groups in the chromosome. Selection maps the cost function $g(s)$ value of each candidate solution (chromosome) to a score function $f(s)$, which is also known as *fitness function*. The function $f(s)$ is a measure of the fitness of each candidate relative to the maximum cost function value g_{max}. The ultimate goal is to maximize the score function or fitness function of each chromosome. An appropriate selection procedure should be designed to select best-performing chromosomes for crossover

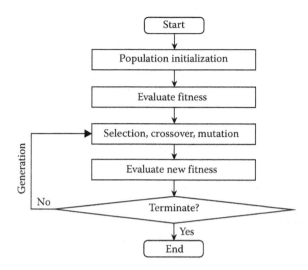

FIGURE 4.1 The basic grouping genetic algorithm.

(Onwubolu and Mutingi, 2001). Therefore, the selection procedure should be able to probabilistically select and store each chromosome s in the mating pool (called *temp*) based on the relative fitness values.

In the crossover procedure, the selected chromosomes are crossed probabilistically to produce a pool of new offspring, called selection pool (*spool*). This allows for exploration of unvisited regions in the solution space (Mutingi and Mbohwa, 2012). Here, the proposed group crossover operator probabilistically exchanges groups of genes of selected pairs of chromosomes, until the desired pool is obtained (Filho and Tiberti, 2006). Further, to intensify local search and to maintain population diversity, mutation is applied to every new chromosome. The goal of mutation is to alter genes in randomly chosen groups of genes in a chromosome.

The iterative loop, comprising selection, crossover, mutation, and evaluation, is executed till a termination criterion is satisfied. However, to infuse more realism into the solution procedure, and to handle the fuzzy imprecise parameters of grouping problems, some of the operators of the GGA procedure should be fuzzified. A fuzzy grouping genetic algorithm is presented to cover this void.

4.3 FUZZY GROUPING GENETIC ALGORITHM

The fuzzy grouping genetic algorithm (FGGA) is a further development of the GGA that incorporates fuzzified procedures—for instance, fuzzy evaluation techniques in fitness evaluation. The FGGA components include chromosome coding and initialization, followed by selection, group crossover, group mutation, and fuzzy evaluation.

4.3.1 FGGA CODING SCHEME

In order to enhance the effectiveness and efficiency of the FGGA, a unique group coding scheme is developed to exploit the group structure of the problem. Figure 4.2

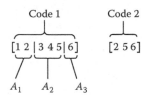

FIGURE 4.2 A representation scheme for grouping problems.

illustrates an example of a coding scheme for grouping problems. The code comprises two codes: code 1 and code 2. Code 1 represents a typical set of three groups—that is, (1,2), (3,4,5), and (6)—separated by the symbol "|," which represents the delimiters. The groups are assigned to the respective assignees A_1, A_2, and A_3. On the other hand, code 2 represents the respective positions of the delimiters or frontiers of the groups.

It is interesting to see that most, if not all, grouping problems can be represented in this form, with little or no adjustment. The aim is to find the best membership for each element or gene in the available groups such that the overall assignment satisfies as much as possible the desired multiple criteria of that particular problem. Therefore, mapping a particular problem to a suitable coding scheme is the most crucial step in the FGGA approach.

4.3.2 Initialization

In initialization, a set or population of initial candidate solutions of the desired size *popsize* are generated for succeeding iterations. The fitness of the initial population may influence the overall computation time of the algorithm. The following procedures may be used to generate good initial population:

1. Random generation of the initial solutions
2. A greedy approach with a guided probabilistic bias
3. Domain-specific constructive heuristics guided by problem constraints

After the initialization phase, FGGA loops through selection, group crossover, group mutation, and evaluation until a termination criterion is satisfied.

4.3.3 Fuzzy Fitness Evaluation

Fuzzy fitness evaluation determines the fitness of each candidate solution (or chromosome) expressed as a function of appropriate fuzzy evaluation functions. The fitness evaluation function captures the imprecise conflicting goals and constraints, providing a measure of the quality of each candidate solution. The evaluation function F_t, at iteration t, is expressed as a function of n normalized functions denoted by μ_h ($h = 1,\ldots,n$). Therefore, F_t may be obtained using fuzzy multifactor evaluation as follows:

$$F_t(s) = \sum_h w_h \mu_h(s) \tag{4.1}$$

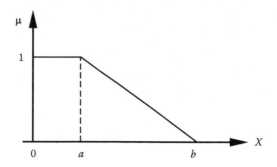

FIGURE 4.3 Linear interval-valued fuzzy membership function.

where s is a candidate solution at iteration t and w_h represents the weight of the function μ_h.

In a fuzzy environment, the quality of a candidate solution represents how much the solution satisfies human preferences and the choices and intuition of the decision maker. To model ambiguity and imprecision, fitness may be calculated using linear interval-valued fuzzy membership functions, such as in Figure 4.3. In this vein, satisfaction level is represented by the most desirable range $[0,a]$ and the decreasing linear function from a to b, where b is the maximum acceptable.

Therefore, the corresponding fuzzy membership function for can be expressed in the following form:

$$\mu_A(x) = \begin{cases} 1 & \text{if } x \le a \\ (b-x)/(b-a) & \text{if } a \le x \le b \\ 0 & \text{if otherwise} \end{cases} \tag{4.2}$$

4.3.4 SELECTION

Selection generally maps a cost function $g(s)$ of a candidate solution s to a fitness function $f(s)$ for evaluation (Onwubolu and Mutingi, 2003). The fitness function of each solution is expressed relative to the maximum cost function value (Goldberg, 1989; Man, Tang, and Kwong, 1999). The aim of the optimization procedure is to maximize fitness $f(s)$ of the form

$$f(s) = \begin{cases} g_{\max} - g(s) & \text{if } g(s) < g_{\max} \\ 0 & \text{if otherwise} \end{cases} \tag{4.3}$$

where $g(s)$ is the cost function of the chromosome and $g_{\max}$ is the largest cost function in the current population. Using the remaining stochastic sampling without replacement (Goldberg, 1989), each chromosome s is selected and stored in the mating pool, *temp*, according to its expected count e_s:

$$e_s = \frac{f(s)}{(1/\text{popsize}) \sum_{s=1}^{\text{popsize}} f(s)} \qquad (4.4)$$

where f_s is the fitness function value of chromosome s. Based on the integer part of e_s, each chromosome receives copies equal to the integer part of e_s denoted by $[e_s]$, while its fractional part is treated as success probability of obtaining additional copies of the same chromosome into the mating pool. In this manner, chromosomes with higher fitness values have a higher probability of surviving into the next generation.

4.3.5 CROSSOVER

Crossover is an evolutionary stochastic mechanism that probabilistically mates selected chromosomes to produce new offspring by exchanging genes between selected pairs of chromosomes at a crossover probability p_c until the desired pool size *poolsize* is obtained. Figure 4.4 illustrates a crossover operation. Groups of genes $|3\ 4\ 5|$ and $|3\ 1|$ are swapped to produce two new offspring.

Oftentimes, the crossover operation may yield offspring with redundant or missing genes. As such, the offspring need to undergo repair. A repair mechanism is incorporated to identify and eliminate redundant genes while adding missing ones. However, crossed genes should not undergo repair. Figure 4.5 shows an example of a repair operation on offspring or a new chromosome [1 2 | 3 1 | 5 6].

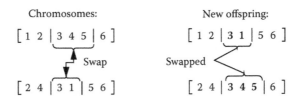

FIGURE 4.4   An example of the crossover operation.

Before repair	:	$[\ 1\ 2\ |\ 3\ 1\ |\ 5\ 6\ ]$
Eliminate 1	:	⇩
		$[\ \ \ 2\ |\ 3\ 1\ |\ 5\ 6\ ]$
Introduce 4	:	⇩
After repair	:	$[\ 2\ 4\ |\ 3\ 1\ |\ 5\ 6\ ]$

FIGURE 4.5   An example of the chromosome repair mechanism.

Chromosome	:	[2 4	3 1	5 6]
Select groups	:	2 and 3
Select and swap genes	:	3 and 6
Mutated chromosome	:	[2 4	6 1	5 3]

FIGURE 4.6 An example of swap mutation.

Offspring	:	[2 4	3 1	5 6]
Frontier = rand(1,2) = 2	:	Shift frontier
Shift direction = rand(R,L) = R	:	Right
Mutated offspring	:	[2 4 3	1	5 6]

FIGURE 4.7 An example of shift mutation.

4.3.6 MUTATION

In order to enhance local search in the neighborhood of candidate solutions in the current population and to maintain population diversity, the algorithm applies two types of mutation operators to every new offspring—namely, *swap mutation* and *shift mutation*. The swap mutation exchanges genes between two randomly chosen groups in a chromosome. Figure 4.6 illustrates the operation of the swap mutation on two new chromosomes, where genes 3 and 6 are randomly selected from groups 2 and 3, and swapped.

After the swap mutation, the shift mutation operator randomly selects a frontier between two adjacent groups and shifts it by one step in either the right (R) or left (L) direction. Figure 4.7 provides an example of shift mutation.

4.3.7 INVERSION AND DIVERSIFICATION

After several iterations, the algorithm is expected to converge to a near-optimal solution. However, to prevent premature convergence to a suboptimal solution, an inversion operator is probabilistically applied to the entire population of candidate solutions. The operator rearranges selected groups of chromosomes in the reverse order (Mutingi and Mbohwa, 2012). Figure 4.8 provides an illustration of the inversion operation.

Before inversion	:	[2 4	3 1	5 6]
After inversion	:	[6 5	1 3	4 2]

FIGURE 4.8 An example of inversion.

The overall FGGA procedure		
1. **Begin**		
2.	Input: FGGA parameters; $t = 0$;	
3.	Initialize population, $P(0)$;	
4.	**Repeat**	
5.		Selection:
6.		Evaluate $P(t)$;
7.		Create temporal population, *tempp*(t);
8.		Group crossover:
9.		Select 2 chromosomes from *tempp*(t);
10.		Apply crossover operator;
11.		Repair offspring if necessary;
12.		Group Mutation:
13.		Mutate $P(t)$;
14.		Add offspring to *newpop*(t);
15.		Replacement:
16.		Compare adjacent *spool*(t) and *oldpop*(t) chromosomes;
17.		Take the ones that fare better;
18.		Select the rest of the strings at a probability;
19.		Diversification:
20.		**While** (diversity level is unacceptable)
21.		diversify $P(t)$;
22.		calculate h;
23.		**End While**
24.		Evaluate $P(t)$;
25.		Advance population:
26.		*oldpop*(t) = *newpop*(t);
27.		Advance population, $t = t + 1$;
28.	**Until** (termination criteria satisfied, e.g., $t \geq T$)	
29. **End**		

FIGURE 4.9　The overall pseudocode for the FGGA procedure.

4.4　THE OVERALL FGGA PROCEDURE

Figure 4.9 summarizes the overall FGGA procedure, incorporating all the operators illustrated in the previous sections. The algorithm accepts genetic probabilities—that is, for crossover, mutation, and diversification operators. Following initialization, the algorithm loops through the genetic operators till a termination criterion is satisfied.

4.5　APPLICATION AREAS

It is realized in this research, that grouping problems can be modeled and solved using the FGGA approach, with little or no adjustment. Some of the problem areas that fall into this category are presented in Table 4.1.

Homecare worker scheduling. Home healthcare scheduling is a multicriteria decision problem concerned with scheduling and routing of healthcare staff to visit patients at their homes (Bachouch et al., 2010; Mutingi and Mbohwa, 2013a). The aim is to satisfy, as much as possible, the quality of the schedule, measured in terms of (1) minimal violation of time windows, (2) maximal workload balance among the

TABLE 4.1

Potential FGGA Application Areas and Their Associated Fuzzy Decision Criteria

No.	Problem Area	Brief Description	Fuzzy Decision Criteria	Selected References
1	Homecare worker scheduling	Scheduling and routing of healthcare workers to visit patients at their homes	To satisfy nurse preferences, workload balance, management choices To satisfy patient expectations, time windows To minimize costs, overstaffing, and understaffing	Begur, Miller, and Weaver (1997); Akjiratikarl, Yenradee, and Drake (2007); Bachouch et al. (2010); Mutingi and Mbohwa (2013a)
2	Care task assignment	Allocation of healthcare tasks to nursing staff so that patients can receive the required healthcare service	To provide a satisfactory service to patients To satisfy management goals To minimize costs, and over/understaffing	Cheng et al. (2007, 2008); Mutingi and Mbohwa (2014)
3	Group technology	Management theory based on the principle that "similar things should be done similarly," e.g., similar activities, tasks, parts	To satisfy customer expectations To satisfy staff preferences and expectations To satisfy management goals and expectations. To minimize costs, transportation	Askin and Standridge (1993); Won and Lee (2001); Mahdavi and Mahadevan (2008); Mutingi et al. (2012)
4	Vehicle routing	Serving clients at their desired time windows with a given fleet of vehicles	To satisfy client preferences, workload balance To satisfy client time window preferences To satisfy management desires and goals, minimize costs, and distance traveled	Teodorovic and Pavkovic (1996); Tarantilis, Kiranoudis, and Vassiliadis (2003, 2004); Braysey and Gendreau (2005); Liu, Huang, and Ma (2009); Tang, Yu, and Li (2015)

assigned staff, and (3) maximal clustering efficiency of each group of patients visited by a single caregiver, so that traveling costs are reduced. As a result, the homecare staff scheduling problem is a complex problem that demands efficient multicriteria optimization methods. The FGGA approach is a potential tool to cover this void.

Care task assignment. Healthcare task assignment is concerned with daily allocation of healthcare tasks to nursing staff so that patients can receive the required healthcare service (Cheng et al., 2007). All tasks, such as assistance with meals, instillation of drops, and preparation of infusions, must be assigned, subject to constraints regarding caregiver capacity, nature of tasks, and their precedence relationships. In hospitals, care tasks are often assigned manually using spreadsheets, based on known patient information on task duration and the time window during which specific tasks should be performed (Cheng et al., 2008). The goal is to provide satisfactory service to patients, to satisfy as much as possible the preferences and choices of nurses, and to satisfy management goals as much as possible. The FGGA approach comes in handy for such a fuzzy multicriteria decision problem (Mutingi and Mbohwa, 2013b).

Other problems. Apart from the problem areas outlined previously, other real-world grouping problems include group technology applications (Won and Lee, 2001; Mutingi et al., 2012), the cutting stock problem (Ramesh, 2001; Onwubolu and Mutingi, 2003), the vehicle routing problem (Tarantilis, Kiranoudis, and Vassiliadis, 2003, 2004; Tang, Yu, and Li, 2015), line balancing (Scholl, 1999; Sabuncuoglu, Erel, and Tanyer, 2000; Scholl and Becker, 2006), and the bin packing problem (Allen, Burke, and Kendall, 2011; Pillay, 2012).

4.6   SUMMARY

Grouping problems are of common occurrence in service and manufacturing industries. As most of these are fuzzy multicriteria decision problems and difficult to solve using conventional methods, it is essential to develop effective and efficient multicriteria decision approaches that can take advantage of the common group structure of the problems. Such approaches are expected to provide a solution approach to a wide range of problems across various disciplines.

This chapter presented a fuzzy grouping genetic algorithm, an improvement from the group genetic algorithm. The approach is motivated by the existence of a number of grouping problems in various disciplines, specifically the healthcare sector. A closer look at the common characteristics of the problems showed that a common group coding scheme can be used to represent most, if not all, of the problems. The proposed FGGA metaheuristic has unique enhanced features, including the group chromosome scheme, group crossover, group mutation, and group chromosome repair mechanism. These group operators enable the algorithm to take advantage of the group structure inherent in the problems.

In practice, the fuzzy multicriteria grouping approach described in this research is handy as it offers a structured way of solving grouping problems with little or no adjustments. Specific problems can easily be mapped into a common FGGA coding scheme. By using fuzzified genetic operators, imprecise parameters of the problems can be modeled with ease. Furthermore, the suggested approach is widely applicable to a number of problem situations. Among the several possible areas of application of the algorithm, the healthcare sector appears to be the most demanding area, given the urgent need to contain healthcare operations costs in nurse scheduling, home-care nurse routing and scheduling, and care task assignment. However, the suggested approach can be extended to a lot more multicriteria decision problems across disciplines in industry.

REFERENCES

Akjiratikarl, C., Yenradee, P. and Drake, P. R. 2007. PSO-based algorithm for home care worker scheduling in the UK. *Computers & Industrial Engineering* 53: 559–583.

Allen, S. D., Burke, E. K. and Kendall, G. 2011. A hybrid placement strategy for the three-dimensional strip packing problem. *European Journal of Operational Research* 209 (1): 219–227.

Askin, R. G. and Standridge, C. R. 1993. *Modeling and analysis of manufacturing systems.* John Wiley & Sons, New York.

Bachouch, R. B., Liesp, A. G., Insa, L. and Hajri-Gabouj, S. 2010. An optimization model for task assignment in home healthcare. *IEEE Workshop on Health Care Management (WHCM),* 1–6.

Begur, S. V., Miller, D. M., and Weaver, J. R. 1997. An integrated spatial DSS for scheduling and routing home-health-care nurses. *Interfaces* 27 (4): 35–48.

Braysey, O. and Gendreau, M. 2005. Vehicle routing problems with time windows, part I: Route construction and local search algorithms. *Transportation Science* 39 (1): 104–118.

Cheng, M., Ozaku, H. I., Kuwahara, N., Kogure, K. and Ota, J. 2007. Nursing care scheduling problem: Analysis of staffing levels. *Proceedings of the 2007 IEEE International Conference on Robotics and Biomimetics,* December 15–18, 2007, Sanya, China, 1: 1715–1719.

Cheng, M., Ozaku, H. I., Kuwahara, N., Kogure, K. and Ota, J. 2008. Simulated annealing algorithm for scheduling problem in daily nursing cares. *Proceedings of the IEEE International Conference on Systems, Man and Cybernetics,* SMC (October), 1681–1687.

Falkenauer, E. 1992. The grouping genetic algorithms—Widening the scope of the GAs. *Belgian Journal of Operations Research, Statistics and Computer Science* 33: 79–102.

Filho, E. V. G. and Tiberti, A. J. 2006. A group genetic algorithm for the machine cell formation problem. *International Journal of Production Economics* 102: 1–21.

Goldberg, D. E. 1989. *Genetic algorithm in search, optimization, and machine learning,* Addison-Wesley, Reading, MA.

Liu, S., Huang, W. and Ma, H. 2009. An effective genetic algorithm for the fleet size and mix vehicle routing problems. *Transportation Research Part E* 45: 434–445.

Mahdavi, I. and Mahadevan, B. 2008. CLASS: An algorithm for cellular manufacturing system and layout design using sequence data. *Robotics and Computer-Integrated Manufacturing* 24: 488–497.

Man, K. F., Tang, K. S. and Kwong, S. 1999. Genetic algorithms: Concepts and design, Springer, London.

Mutingi, M. and Mbohwa, C. 2012. Enhanced group genetic algorithm for the heterogeneous fixed fleet vehicle routing problem. *IEEE IEEM Conference on Industrial Engineering and Engineering Management,* Hong Kong, 207–211.

Mutingi, M. and Mbohwa, C. 2013a. Home healthcare worker scheduling: A group genetic algorithm approach. *Proceedings of World Congress on Engineering,* London, (July), 721–725.

Mutingi, M. and Mbohwa, C. 2013b. Task assignment in home health care: A fuzzy group genetic algorithm approach. *The 25th Annual Conference of the Southern African Institute of Industrial Engineering* 2013, Stellenbosch, South Africa, 9–11 July.

Mutingi, M. and Mbohwa, C. 2014. A fuzzy grouping genetic algorithm for care task assign-ment. *IAENG International Conference on Systems Engineering and Engineering Management*, 2014, San Francisco (October), 454–459.

Mutingi, M., Mbohwa, C., Mhlanga, S. and Goriwondo, W. 2012. Integrated cellular manu-facturing system design: An evolutionary algorithm approach. *Proceedings of the 3rd International Conference on Industrial Engineering and Operations Management*, Turkey (July), 254–264.

Mutingi, M. and Onwubolu, G. C. 2012. Integrated cellular manufacturing system design and layout using group genetic algorithms. In *Manufacturing system*, ed. Faieza Abdul Aziz. InTech-Open (http://www.intechopen.com/books/manufacturing-system), 205–222.

Onwubolu, G. C. and Mutingi, M. 2001. A genetic algorithm approach to cellular manufactur-ing systems. *Computers & Industrial Engineering* 39: 125–144.

Onwubolu, G. C. and Mutingi, M. 2003. A genetic algorithm approach for the cutting stock problem. *Journal of Intelligent Manufacturing* 14: 209–218.

Pillay, N. 2012. A study of evolutionary algorithm selection hyper-heuristics for the one-dimensional binpacking problem. *South African Computer Journal* 48: 31–40.

Ramesh, R. 2001. A generic approach for nesting of 2-D parts in 2-D sheets using genetic and heuristic algorithms. *Computer-Aided Design* 33 (12): 879–891.

Sabuncuoglu, I., Erel, E. and Tanyer, M. 2000. Assembly line balancing using genetic algo-rithms. *Journal of Intelligent Manufacturing* 11 (3): 295–310.

Scholl, A. 1999. *Balancing and sequencing of assembly lines*. Physica-Verlag, Heidelberg.

Scholl, A. and Becker, C. 2006. State-of-the-art exact and heuristic solution procedures for simple assembly line balancing. *European Journal of Operational Research* 168: 666–693.

Tang, J., Yu, Y. and Li, J. 2015. An exact algorithm for the multi-trip vehicle routing and scheduling problem of pickup and delivery of customers to the airport. *Transportation Research Part E: Logistics and Transportation Review* 73: 114–132.

Taillard, E. D. 1999. A heuristic column generation method for the heterogeneous fleet VRP. *RAIRO* 33: 1–34.

Tarantilis, C. D., Kiranoudis, C. T. and Vassiliadis, V. S. A. 2003. A list based threshold accept-ing metaheuristic for the heterogeneous fixes fleet vehicle routing problem. *Journal of the Operational Research Society* 54 (1): 65–71.

Tarantilis, C. D., Kiranoudis, C. T. and Vassiliadis, V. S. A. 2004. A threshold accepting meta-heuristic for the heterogeneous fixed fleet vehicle routing problem. *European Journal of Operations Research* 152: 148–158.

Teodorovic, D. and Pavkovic, G. 1996. The fuzzy set theory approach to the vehicle routing problem when demand at nodes is uncertain. *Fuzzy Sets and Systems* 82 (3): 307–317.

Toth, P. and Vigo, D. 2002. The vehicle routing problem. SIAM monograph on discrete math-ematics and applications. SIAM, Philadelphia, PA.

Won, Y. and Lee, K. C. 2001. Group technology cell formation considering operation sequences and production volumes. *International Journal of Production Research* 39: 2755–2768.

5 Fuzzy Grouping Particle Swarm Optimization

5.1 INTRODUCTION

Healthcare and other service systems are characterized with decision problems that involve grouping of items or activities in order to improve the operational efficiency and effectiveness of that system. For example, when assigning tasks to nurses in a hospital ward, the decision maker has to allocate a group of tasks to each nurse so that the overall schedule satisfies the desires and expectations of patients, nurses, and management (Mutingi and Mbohwa, 2014). As such, it is quite crucial for the decision maker to know how to efficiently create groups of tasks and to assign them to workers as satisfactorily as possible.

The task assignment in a home healthcare setting is similar to care task assignment in a hospital ward. Groups of homecare tasks are usually assigned to a set of available healthcare workers, where each task is defined by its task duration, and its earliest and latest start times (Bachouch et al., 2010). More often than not, combined routing and scheduling decisions are involved (Begur, Miller, and Weaver, 1997; Hertz and Lahrichi, 2009; Mutingi and Mbohwa, 2013a). Typical tasks are usually in the form of patient visits, administrative duties, and drug delivery, among others (Mutingi and Mbohwa, 2014). These and other related problem situations, such as vehicle routing problems (Braysey and Gendreau, 2005), group technology problems (Won and Lee, 2001), and cell manufacturing system design (Filho and Tiberti, 2006; Scholl and Becker, 2006) are best described as *grouping problems*.

In all the preceding and similar cases, high-quality decisions such as task assignments, patient assignments, care task schedules, and homecare staff schedules must be coordinated to satisfy the expectations and wishes of patients, staff, and management. Thus, the problems have to be viewed from a multiple criteria point of view. However, when considering subjective criteria like satisfaction of workload balance and task time window preferences, nonstochastic uncertainties or imprecision is involved. In other words, the goals associated with workload fairness and patient satisfaction are in most cases imprecise, or fuzzy. As a result of their complex features, grouping problems are hard to solve. Notable complicating features of grouping problems are outlined as follows:

1. Grouping problems are highly combinatorial in nature
2. Grouping problems are highly constrained
3. The problems have imprecise parameters arising from human preferences and choices that are difficult to quantify

Due to their complicating features, grouping problems lend themselves to meta-heuristic algorithms and expert systems. Some potential metaheuristic methods are particle swarm optimization, tabu search, genetic algorithms, and evolutionary algorithms. However, these methods have to be enhanced and designed so that they take advantage of the group structure of the problems. This chapter focuses on the development of a fuzzy grouping particle swarm optimization algorithm for solving grouping problems. The objectives of the research are as follows:

1. To describe the basic particle swarm optimization algorithm
2. To develop a fuzzy grouping particle swarm optimization algorithm as an extension of the basic particle swarm optimization algorithm
3. To outline the potential application areas of the proposed algorithm

The rest of the chapter is structured as follows: The next section presents a brief background to particle swarm optimization. This is followed by the proposed fuzzy grouping particle swarm optimization. Potential application areas of the algorithm are then outlined. The last section summarizes the chapter.

5.2   PARTICLE SWARM OPTIMIZATION

Particle swarm optimization (PSO), first introduced by Kennedy and Eberhart (1995), is a stochastic parallel population-based optimization technique inspired by the social behavior of fish schooling and bird flocking. Each bird or fish, called particle, is treated as a point in a d-dimension space, and the status of a particle is characterized by its position and velocity. The mechanics of the algorithm makes use of a velocity vector to update the current position of each particle in the swarm. While flying, each particle adjusts its position according to its own experience and that of the most successful particle in the swarm (Shi and Eberhart, 1998). The velocity and the position updates of each particle are respectively determined by the following expressions:

$$v_i(t + 1) = v_i(t) + c_1 \cdot \eta_1 \cdot (pbest_i(t) - x_i(t)) + c_2 \cdot \eta_2 \cdot (gbest(t) - x_i(t)) \qquad (5.1)$$

$$x_i(t + 1) = x_i(t) + v_i(t + 1) \qquad (5.2)$$

where
  $v_i(t)$ and $x_i(t)$ are the velocity component and the location component of particle i
      at iteration t, respectively
  $v_i(t + 1)$ and $x_i(t + 1)$ are the velocity component and the location component of
      particle i at iteration $t + 1$, respectively
  $pbest_i$ is the best location of particle i
  $gbest_i$ is the global best location of the whole swarm
  c_1 and c_2 are the cognitive and social parameters, respectively
  η_1 and η_2 are uniformly generated random numbers in the range [0,1]

Figure 5.1 shows the basic flowchart of the PSO procedure. After population initialization, the algorithm calculates the fitness of each particle according to a given

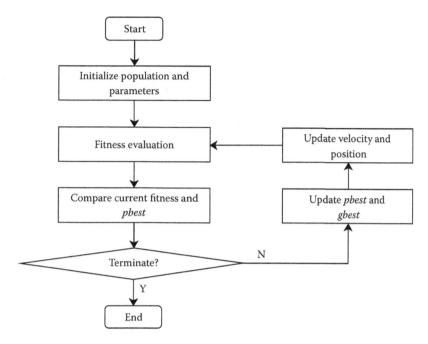

FIGURE 5.1 Basic PSO flowchart.

fitness function. The current fitness of each particle is compared with its previous fitness in order to update each particle's *pbest*. If the termination criteria are satisfied, the best *pbest* is usually the selected solution. Otherwise, *pbest* and *gbest* are updated, and the velocity and position of each particle are then updated accordingly. As depicted in Figure 5.1, fitness evaluation is repeated, closing the PSO iterative loop.

5.3 FUZZY GROUPING PARTICLE SWARM OPTIMIZATION

Fuzzy grouping particle swarm optimization (FGPSO) is a development from the basic PSO algorithm. It incorporates unique group coding schemes and fuzzified mechanisms such as fuzzy evaluation. Its grouping schemes are designed to exploit the group characteristics common in grouping problems, so as to enhance the optimization search process.

FGPSO initially accepts input parameters and then randomly generates an initial swarm of particles. The generated particles then fly at a certain velocity, so as to find a global best position after a number of iterations. As in the basic PSO procedure, each particle iteratively adjusts its velocity according to its momentum, its best position (*pbest*), and that of its neighbors (*gbest*). This eventually determines its new position. Assuming a search space of dimension d, and a total number of particles n, then the position of the ith particle is $x_i = [x_{i1}, x_{i2},...,x_{id}]$, its best position is $pbest_i = [pbest_{i1}, pbest_{i2},...,pbest_{id}]$, and its velocity is $v_i = [v_{i1}, v_{i2},...,v_{id}]$. It follows that the

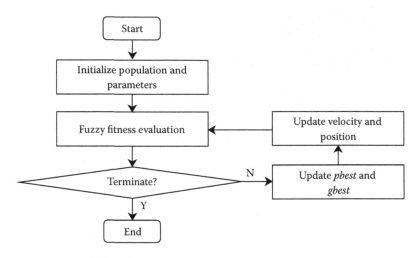

FIGURE 5.2   A flowchart of the proposed FGPSO.

position and velocity of the particle at iteration $(t + 1)$ are updated according to the following expressions:

$$v_i(t + 1) = w \cdot v_i(t) + c_1 \cdot \eta_1 \cdot (pbest_i(t) - x_i(t)) + c_2 \cdot \eta_2 \cdot (gbest(t) - x_i(t)) \quad (5.3)$$

$$x_i(t + 1) = x_i(t) + v_i(t + 1) \quad (5.4)$$

where
  c_1 and c_2 are constants
  η_1 and η_2 are uniformly distributed random variables in the range [0,1]
  w is an inertia weight showing the effect of the previous velocity on the new
    velocity vector

Figure 5.2 shows a flowchart summarizing the logic of the FGPSO approach, consisting of initialization, particle coding scheme, fitness evaluation, and velocity update.

5.3.1   GROUP CODING SCHEME

To exploit the group structure of the grouping problems, the proposed FGPSO algorithm uses a group coding scheme. For instance, consider a decision maker who is supposed to assign a set of six tasks to three healthcare staff or caregivers. In this connection, let $A = [1, 2, 3,...,6]$ represent a string of six tasks to be performed by the three caregivers. Task assignment involves partitioning tasks along A into three groups so that desired objectives are satisfied—for example, (1) minimize workload variation as much as possible, and (2) satisfy as much as possible, the desired time to execute each task. Figure 5.3 shows the structure of the coding scheme for the problem. Part 1 of the code represents the actual allocation of the tasks. Tasks 1 and 3

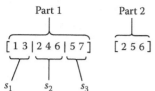

FIGURE 5.3 An FGPSO group coding scheme.

are assigned to staff s_1, tasks 2, 4, and 6 are assigned to s_2, while tasks 5 and 7 are assigned to s_3. On the other hand, part 2 records the positions of the delimiters separating the task groups [1 3], [2 4], and [5].

5.3.2 INITIALIZATION

Using a suitable grouping code as presented in the previous section, an initial population of candidate solutions is generated. A number of heuristic procedures may be used to generate good initial solutions, for instance:

1. Domain-specific constructive heuristics guided by problem constraints
2. Approach with a guided probabilistic bias
3. Random generation of the initial solutions

When generating candidate solutions, either real-valued or integer-coded schemes are used, depending on the nature of the problem. Unique problem situations may require a mix of real and integer-coded values. Continuous variables are coded into real-valued positions using the following expression:

$$x_i = X_{min} + (X_{max} - X_{min}) \times U(0,1) \tag{5.5}$$

where X_{min} and X_{max} are the lower and upper limits of a predefined range of position values, respectively, and $U(0,1)$ is a uniform random number in the range [0,1].

To generate integer-valued codes, continuous position values are converted into integer positions. The generated position values are mapped to the nearest integer number using a suitable rounding function, $round()$, as follows:

$$x_i = X_{min} + round((X_{max} - X_{min}) \times U(0,1)) \tag{5.6}$$

Following initialization, FGPSO loops through fuzzy evaluation, velocity, and position update until a termination criterion is satisfied.

5.3.3 FUZZY EVALUATION

In a fuzzy environment, the quality of a solution represents how much the solution satisfies human preferences and the choices and intuition of the decision maker. Fuzzy fitness evaluation uses fuzzy evaluation techniques to determine the fitness of each particle or candidate solution in the swarm. This is achieved by using fitness

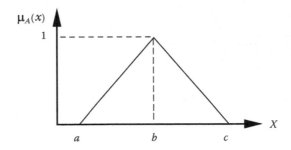

FIGURE 5.4 A triangular fuzzy membership function (a,b,c).

evaluation functions that can capture the imprecise conflicting multiple criteria, such as management goals and decision maker preferences and choices, in order to provide a measure of the quality of each particle. Let F_t denote the evaluation function at iteration t. Assume that F_t can be expressed as a function of m normalized functions, μ_h ($h = 1,...,m$). It follows that F_t can be obtained using the following expression:

$$F_t(i) = \sum_h w_h \mu_h(i) \qquad (5.7)$$

where w_h represents the weight of each function μ_h ($h = 1,...,m$) for particle or candidate solution i at iteration t.

To model ambiguity and imprecision, linear interval-valued fuzzy membership functions may be used to represent satisfaction levels of the decision maker's choices and preferences. In practice, triangular fuzzy numbers are used for this purpose. A triangular fuzzy parameter is generally defined by $A = <a,b,c>$, where b is the most desired value of the fuzzy parameter, and a and c are the minimum and maximum acceptable values, respectively. Figure 5.4 shows a triangular fuzzy membership function

The membership function of x in the fuzzy set A, $\mu_A: X \rightarrow [0,1]$, is given by the following expression:

$$\mu_i(x) = \begin{cases} (x-a)/(b-a) & \text{if } a \leq x \leq b \\ (x-b)/(c-b) & \text{if } b \leq x \leq c \\ 0 & \text{if otherwise} \end{cases} \qquad (5.8)$$

5.3.4 Velocity and Position Update

As in the PSO mechanism, FGPSO makes use of a velocity vector to update the current position of each particle in the swarm. The velocity and the position updates of each particle at iteration $(t + 1)$ are respectively determined by the following expressions:

$$v_i(t + 1) = w \cdot v_i(t) + c_1 \cdot \eta_1 \cdot (pbest_i(t) - x_i(t)) + c_2 \cdot \eta_2 \cdot (gbest(t) - x_i(t)) \quad (5.9)$$

$$x_i(t + 1) = x_i(t) + v_i(t + 1) \quad (5.10)$$

where

w is an inertia weight showing the effect of previous velocity on the new velocity vector

c_1 and c_2 are constants

η_1 and η_2 are uniformly distributed random variables in $[0,1]$

5.4   THE OVERALL FGPSO ALGORITHM

The previous sections described the procedures of operators of the FGPSO algorithm. Deriving from these procedures, the overall structure of the FGPSO algorithm is presented in the form of a pseudocode in Figure 5.5.

Fuzzy grouping particle swarm optimization algorithm

1. Input $w, \eta_1, \eta_2, c_1, c_2, N$;
2. Initialization;
3. For $i = 1$ to N:
4.   Initialize particle position $x_i(0)$ and velocity $v_i(0)$;
5.   Initialize $pbest_i(0)$;
6. **End For**
7. Initialize $gbest(0)$;
8. For $i = 1$ to N:
9.   Compute fuzzy fitness $f(x)$, $x = (x_1, x_2,...,x_N)$;
10. **Repeat**
11.   For $i = 1$ to N:
12.   Compute fuzzy fitness f_i;
13.   If ($f_i >$ current $pbest$) **Then**
14.   Set current value as new $pbest$;
15.   If ($f_i >$ current $gbest$) **Then**
16. $gbest = i$; **End If**
17.   **End If**
18.   **End For**;
19.   For $i = 1$ to N:
20.   Find neighborhood best;
21.   Compute particle velocity $v_i(t + 1)$;
22.   Update particle position $x_i(t + 1)$;
23.   **End For**;
24. **Until** (Termination criteria are satisfied);

FIGURE 5.5   A pseudocode for the FGPSO algorithm.

The algorithm accepts input of the parameters w, η_1, η_2, c_1, c_2, and N from the user. A population of N candidate solutions is generated at random. Depending on the nature of the variables of the problem, real- or integer-valued or candidate solutions are generated based on the group coding scheme.

The proposed FGPSO algorithm has a number of desirable advantages in its application. First, its procedure is easy to follow and therefore can be implemented easily in a number of problem situations. It is robust and versatile in that it is applicable to similar problems with little or no fine-tuning. Moreover, the algorithm is computationally efficient, obtaining good solutions within reasonable computation times. Most importantly, incorporating fuzzy evaluation into the algorithm allows the global optimization process to pass through inferior solutions, which will eventually yield improved solutions. Fuzzy multicriteria evaluation ensures that instances of infeasible solutions are avoided during algorithm execution.

A number of problems can be modeled and solved using the proposed FGPSO algorithm. Examples of potential application areas of the proposed algorithm are presented in the next section.

5.5 POTENTIAL APPLICATION AREAS

It can be seen throughout the chapter that various types of grouping problems lend themselves to the FGPSO approach, with little or no adjustments. To illustrate, it is important to present some of the problem areas falling into this category, as shown in Table 5.1.

5.5.1 CARE TASK ASSIGNMENT IN HOSPITALS

In a hospital setting, healthcare task assignment seeks to fairly allocate healthcare tasks, such as instillation of drops, assistance with meals, and preparation of infusions, to nursing staff on a daily basis. The primary aim is to ensure that patients receive the expected healthcare service in a timely manner (Mutingi and Mbohwa, 2014). Thus, the goal is to satisfy as much as possible the wishes and expectations of patients, nursing staff, and management. All tasks are assigned consideration of caregiver capacity limitations, the nature of tasks, and their precedence relationships (Cheng et al., 2007, 2008). Without enhanced multicriteria decision methods such as FGPSO, decision makers in hospitals often assign care tasks manually using spreadsheets, patient information on task duration, and the time window during which specific tasks should be performed (Cheng et al., 2008). The FGPSO algorithm is a potential approach for modeling and solving fuzzy multicriteria decision problems in the healthcare sector (Mutingi and Mbohwa, 2014).

5.5.2 HOMECARE WORKER SCHEDULING

Decision makers in home healthcare service organizations are concerned with coordinating patient visits at their homes. This is a multicriteria decision problem involving scheduling and routing of healthcare workers so that patients are visited

TABLE 5.1

Potential FGPSO Application Areas and Their Fuzzy Multiple Decision Criteria

No.	Problem Area	Brief Description	Fuzzy Decision Criteria	Selected References
1	Care task assignment in hospitals	Allocation of healthcare tasks to nurses to ensure that patients receive the necessary healthcare service	To provide a satisfactory service to patients To construct a high-quality schedule that satisfies nurses To satisfy management goals To minimize costs and over/understaffing	Cheng et al. (2007, 2008); Mutingi and Mbohwa (2014)
2	Home healthcare worker scheduling	Scheduling and routing of healthcare workers to visit patients at their homes	To satisfy nurse preferences, expectations, and time windows To produce quality schedules that balance workload allocation, satisfying workers To satisfy management goals and choices; minimize costs	Akjiratikarl, Yenradee, and Drake (2007); Bachouch et al. (2010); Mutingi and Mbohwa (2013)
3	Vehicle routing problem	Serving clients at their desired time windows with a given fleet of vehicles	To satisfy client preferences, workload balance To satisfy client time window preferences To satisfy management desires and goals, minimize costs, and distance traveled	Teodorovic and Pavkovic (1996); Tarantilis, Kiranoudis, and Vassiliadis (2004); Tang et al. (2009)
4	Group technology	Management theory based on the principle that "similar things should be done similarly" (e.g., similar activities, tasks, parts)	To satisfy customer expectations To satisfy staff preferences and expectations To satisfy management goals and expectations To minimize transportation costs	Selim, Askin, and Vakharia (1998); Won and Lee (2001); Mahdavi and Mahadevan (2008); Mutingi and Onwubolu (2012)

during their preferred time windows (Bachouch et al., 2010; Mutingi and Mbohwa, 2013a,b). The desired quality of schedules must be satisfied as much as possible. Schedule quality may be measured based on such criteria as (1) violation of preferred time windows, (2) workload balance among the assigned staff, (3) clustering efficiency of each group of patients visited by a single caregiver, and (4) traveling costs. In this regard, efficient multicriteria decision methods such as FGPSO may be more appropriate and effective.

5.5.3 OTHER PROBLEMS

The grouping concepts can be extended to other problems in the healthcare sector and elsewhere. Similar grouping problems in other areas include group technology applications (Won and Lee, 2001), manufacturing systems layout (Filho and Tiberti, 2006; Mutingi and Onwubolu, 2012), balancing and sequencing of assembly lines (Scholl, 1999; Sabuncuoglu, Erel, and Tanyer, 2000), the cutting stock problem (Ramesh, 2001; Onwubolu and Mutingi, 2003), the vehicle routing problem (Tarantilis, Kiranoudis, and Vassiliadis, 2003, 2004; Tang et al., 2009), line balancing (Scholl, 1999; Scholl and Becker, 2006), and the bin packing problem (Pillay, 2012).

5.6   SUMMARY

Grouping problems are usually associated with multiple criteria decisions that are often imprecise and conflicting in nature. As such, developing fuzzy multicriteria decision support tools is imperative. For instance, in a homecare setting, workload must be balanced among healthcare workers, patients must be visited at their preferred time windows, and management goals must be satisfied to the highest degree possible. Where decision criteria are ill-defined or imprecise, the use of fuzzy set theory concepts is beneficial.

   This chapter presented a unique FGPSO approach that can model fuzzy grouping problems such as care task assignment in a hospital ward, homecare worker scheduling, homecare task assignment, and other related problems in manufacturing and service industries. The suggested heuristic approach provides useful contributions to researchers, academicians, and practitioners, especially in healthcare services. It provides a potential approach to solving staff scheduling problems when the desired management goals and worker preferences are imprecise or ill-structured. Incorporating fuzzy set theory concepts brings more realism to the solution process. Moreover, the algorithm is capable of handling large-scale problems and providing useful solutions in a reasonable computation time, unlike linear programming methods. In this context, the proposed solution method is an invaluable approach that can be developed further into a decision support system for healthcare staff scheduling and related problems.

   To the decision maker in practice, the proposed multicriteria decision approach provides an opportunity to use weights by which the decision maker can incorporate his or her preferences and choices in an interactive manner. Practicing decision makers appreciate the use of interactive decision support rather than the use of methods that prescribe a single solution. Population-based heuristics provide a list of good alternative solutions from which a decision maker can choose the most appropriate solution, considering other practical factors. Overall, FGPSO is a potentially effective solution method that can be used for solving healthcare staff scheduling problems as well as other related problems with group structures.

REFERENCES

Akjiratikarl, C., Yenradee, P. and Drake, P. R. 2007. PSO-based algorithm for home care worker scheduling in the UK. *Computers & Industrial Engineering* 53: 559–583.

Bachouch, R. B., Liesp, A. G., Insa, L. and Hajri-Gabouj, S. 2010. An optimization model for task assignment in home healthcare. *IEEE Workshop on Health Care Management (WHCM)*, 1–6.

Begur, S. V., Miller, D. M. and Weaver, J. R. 1997. An integrated spatial decision support system for scheduling and routing home health care nurses. Technical report, Institute of Operations Research and the Management Science.

Braysey, O. and Gendreau, M. 2005. Vehicle routing problems with time windows, part I: Route construction and local search algorithms. *Transportation Science* 39 (1): 104–118.

Cheng, M., Ozaku, H. I., Kuwahara, N., Kogure, K. and Ota, J. 2007. Nursing care scheduling problem: Analysis of staffing levels. *Proceedings of the 2007 IEEE International Conference on Robotics and Biomimetics*, December 15–18, 2007, Sanya, China, 1: 1715–1719.

Cheng, M., Ozaku, H. I., Kuwahara, N., Kogure, K. and Ota, J. 2008. Simulated annealing algorithm for scheduling problem in daily nursing cares. *Proceedings of the IEEE International Conference on Systems, Man and Cybernetics*, SMC (October), 1681–1687.

Filho, E. V. G. and Tiberti, A. J. 2006. A group genetic algorithm for the machine cell formation problem. *International Journal of Production Economics* 102: 1–21.

Hertz, A. and Lahrichi, N. 2009. A patient assignment algorithm for home care services. *Journal of the Operational Research Society* 60 (4): 481–495.

Kennedy, J. and Eberhart, R. C. 1995. Particle swarm optimization. *IEEE International Conference on Neural Networks* 4: 1942–1948.

Mahdavi, I. and Mahadevan, B. 2008. CLASS: An algorithm for cellular manufacturing system and layout design using sequence data. *Robotics and Computer-Integrated Manufacturing* 24: 488–497.

Mutingi, M. and Mbohwa, C. 2013a. Home healthcare worker scheduling: A group genetic algorithm approach. In *Lecture Notes in Engineering and Computer Science: Proceedings of the World Congress on Engineering*, ed. S. I. Ao, L. Gelman, D. W. L. Hukins, A. Hunter and A. M. Korsunsky, 3–5 July, London, 721–725.

Mutingi, M. and Mbohwa, C. 2013b. Task assignment in home health care: A fuzzy group genetic algorithm approach. *The 25th Annual Conference of the Southern African Institute of Industrial Engineering* 2013, Stellenbosch, South Africa, 9–11 July.

Mutingi, M. and Mbohwa, C. 2014. A fuzzy grouping genetic algorithm for care task assignment. *IAENG International Conference on Systems Engineering and Engineering Management*, 2014, San Francisco (October), 454–459.

Mutingi, M. and Onwubolu, G. C. 2012. Integrated cellular manufacturing system design and layout using group genetic algorithms. In *Manufacturing system*, ed. Faieza Abdul Aziz. InTech-Open (http://www.intechopen.com/books/manufacturing-system), 205–222.

Onwubolu, G. C. and Mutingi, M. 2003. A genetic algorithm approach for the cutting stock problem. *Journal of Intelligent Manufacturing* 14: 209–218.

Pillay, N. 2012. A study of evolutionary algorithm selection hyper-heuristics for the one dimensional bin packing problem. *South African Computer Journal* 48: 31–40.

Ramesh, R. 2001. A generic approach for nesting of 2-D parts in 2-D sheets using genetic and heuristic algorithms. *Computer-Aided Design* 33 (12): 879–891.

Sabuncuoglu, I., Erel, E. and Tanyer, M. 2000. Assembly line balancing using genetic algorithms. *Journal of Intelligent Manufacturing* 11 (3): 295–310.

Scholl, A. 1999. *Balancing and sequencing of assembly lines*. Physica-Verlag, Heidelberg.

Scholl, A. and Becker, C. 2006. State-of-the-art exact and heuristic solution procedures for simple assembly line balancing. *European Journal of Operational Research*, 168: 666–693.

Selim, H. M., Askin, R. G., and Vakharia, A. J. 1998. Cell formation in group technology: Evaluation and directions for future research. *Computers & Industrial Engineering* 34 (1): 3–20.

Shi, Y. and Eberhart R. C. 1998. A modified particle swarm optimizer. *IEEE International Conference on Evolutionary Computation*, 69–73.

Tang, J., Pan, Z., Fung, R. Y. K. and Lau, H. 2009. Vehicle routing problem with fuzzy time windows. *Fuzzy Sets and Systems* 160: 683–695.

Tarantilis, C. D., Kiranoudis, C. T. and Vassiliadis, V. S. A. 2003. A list based threshold accepting metaheuristic for the heterogeneous fixes fleet vehicle routing problem. *Journal of the Operational Research Society* 54 (1): 65–71.

Tarantilis, C. D., Kiranoudis, C. T. and Vassiliadis, V. S. A. 2004. A threshold accepting metaheuristic for the heterogeneous fixed fleet vehicle routing problem. *European Journal of Operations Research* 152: 148–158.

Teodorovic, D. and Pavkovic, G. 1996. The fuzzy set theory approach to the vehicle routing problem when demand at nodes is uncertain. *Fuzzy Sets and Systems* 82 (3): 307–317.

Won, Y. and Lee, K. C. 2001. Group technology cell formation considering operation sequences and production volumes. *International Journal of Production Research* 39: 2755–2768.

Section III

Research Applications

6 Fuzzy Simulated Metamorphosis Algorithm for Nurse Scheduling

6.1 INTRODUCTION

Nurse scheduling is concerned with the assignment of work shifts and off days to nurses over a planning horizon of about 1 week to 1 month. This is the most common challenge in hospitals. The desire is to produce high-quality work schedules that satisfy (1) expectations of patients regarding quality of service, (2) nurse preferences such as workload, and (3) management goals. In the real world, these desires are usually conflicting, imprecise, and uncertain in a nonstochastic sense. As a result, developing high-quality nurse schedules is difficult in such a fuzzy environment. This decision problem situation is commonplace in healthcare organizations (Jan, Yamamoto, and Ohuchi, 2000; Topaloglu and Selim, 2010).

In a fuzzy environment, nurse scheduling, or nurse rostering, is a complex task due to the presence of fuzzy multiple criteria. The problems require interactive tools that are fast, flexible, and adaptable to various problem situations. Decision makers frequently desire to use judicious approaches that can find a trade-off between the many goals, from a multicriteria viewpoint (Vasant and Barsoum, 2006; Villacorta, Masegosa, and Lamata, 2013). Addressing uncertainties, ambiguity, or imprecision in the desired goals is quite appropriate for practicing decision makers (Mutingi and Mbohwa, 2014a). For example, in hospitals, nurses are allowed to express their individual preferences on shift schedules. Patient preferences and expectations should be considered as well. The decision maker must incorporate preferences, choices, and management goals (Topaloglu and Selim, 2010). For shift fairness and equity among the nursing staff, nurse workload should be balanced. These imprecise and conflicting factors have to be considered when constructing work schedules (Inoue et al., 2003). Multicriteria metaheuristics are a viable option—specifically, nature-inspired approaches (Otero, Masegosa, and Terrazas, 2014).

Considering the preceding issues, this chapter presents a novel fuzzy simulated metamorphosis algorithm. The algorithm is inspired by the concepts of biological metamorphosis evolution (Mutingi and Mbohwa, 2014b). The algorithm is motivated by the need for fuzzy, multicriteria, and fast optimization approaches to solving problems with fuzzy goals, preferences, and constraints. In particular, the specific objectives are as follows:

1. To present the nurse scheduling problem in a hospital setting
2. To present a multicriteria fuzzy evolutionary algorithm deriving from biological metamorphosis
3. To apply the algorithm to nurse scheduling problems, demonstrating its efficiency and effectiveness

The rest of the chapter is structured as follows. The next section presents the nurse scheduling problem. A fuzzy simulated metamorphosis is then presented. Thereafter, computational experiments and analyses are provided. The chapter ends with a summary pointing out the strengths of the algorithm and further research prospects.

6.2   NURSE SCHEDULING PROBLEM

The nurse scheduling problem (NSP) or nurse rostering problem (NRP) is a multicriteria decision problem that involves assignment of different types of shifts and off days to nurses over a period of 1 week to 1 month (Shaffer, 1991; Mutingi and Mbohwa, 2013). In practice, contractual work agreements govern the number of assignable shifts and off days per week (Burke, De Causmaecker, and Vanden Berghe, 2004). Imprecise personal preferences should be satisfied as much as possible. Nurses are usually entitled to night shift "e," late night shift "n," and day shift "d," with holidays or days-off "o" (Cheang et al., 2003; Musliu, 2006; Brucker, Burke, and Curtois, 2010). Table 6.1 lists and describes common shift types and their time allocations.

Imprecise nurse and patient preferences should be satisfied to the highest degree possible; that is, the higher the degree of satisfaction is, the higher the schedule quality will be (Mutingi and Mbohwa, 2013). This ensures healthcare service quality, as well as a satisfactory healthcare work environment (job satisfaction), subject to various constraints.

A review of various case studies shows that NSP constraints can be classified into sequence, schedule, and roster constraints as listed in Table 6.2. Sequence constraints relate to the successive order of shifts in an individual nurse schedule or shift pattern. On the other hand, schedule constraints impose restrictions, over the planning period, on each nurse schedule, based on criteria such as workload and number of night shifts. In addition, a roster constraint controls the final combination of nurse schedules, called roster, based on suitable criteria such as shift coverage.

TABLE 6.1
Typical Shift Types

Shift Type	Shift Description	Time Slot
d	Day shift	0800 to 1600 hours
e	Night shift	1600 to 2400 hours
n	Late night shift	0000 to 0800 hours
o	Off day or holiday	0000 to 2400 hours

TABLE 6.2
Constraint Categories

Constraint Category	Constraint Description
A. Sequence	1. Shift sequences n-d, e-n, and e-d are not permissible
	2. Minimum rest time between night shift n
	3. Maximum and minimum number of working days or working hours
B. Schedule	4. Fair or equal total workload assignment
	5. Interval between night shifts should be ≥ 1 week
	6. Fair number of requested days off or holiday assigned
C. Roster	7. Shift coverage requirements to fulfil service quality
	8. Tutorship, where a trainer has to work with a specific trainee
	9. Congeniality, where two or more workmates are not compatible

6.3 FUZZY SIMULATED METAMORPHOSIS FOR NURSE SCHEDULING

This section presents an application of fuzzy simulated metamorphosis (FSM) on an NSP in a fuzzy environment with multiple criteria.

6.3.1 FSM CODING SCHEME

A unique coding scheme is proposed to enhance the FSM performance. Assume that nine nurses are to be assigned evening shift "e," day shift "d," night shift "n," or day-off shift "o." Furthermore, assume that a demand of two nurses is required for shifts e, d, and n, over a planning period of 7 days. The FSM coding scheme allocates one of the four shifts in each day, subject to shift, schedule, and roster constraints. Figure 6.1 presents the FSM code for this example.

Nurse	Mon	Tue	Wed	Thu	Fri	Sat	Sun	d	e	n
Nurse 1	d	d	d	d	d	o	o	5	0	0
Nurse 2	o	d	d	d	d	d	d	6	0	0
Nurse 3	d	o	o	n	n	n	n	1	4	0
Nurse 4	o	o	o	o	e	e	e	0	3	0
Nurse 5	e	e	e	e	o	o	o	0	4	0
Nurse 6	n	n	n	n	n	o	o	0	0	5
Nurse 7	o	o	e	e	e	e	e	0	5	0
Nurse 8	e	e	o	o	e	e	e	0	5	0
Nurse 9	n	n	n	o	o	d	d	2	0	3
d	2	2	2	2	2	2	2			
e	2	2	2	2	2	2	2			
n	2	2	2	2	2	2	2			

(Days span Mon–Sun)

FIGURE 6.1 An example of an FSM coding scheme for nurse scheduling.

FSM initialization
1.   Initialize, counter $i = 1$;
2.   **Repeat**
3.   Initialize $k = 1$;
4.   Randomly generate an initial shift s_1;
5.   **Repeat**
6.   Select shift $s_{k+1} = \text{rand}(d, e, n, o)$ with a probability;
7.   If sequence $s = [s_k s_{k+1}]$ $\in$ Forbidden set F, **Then**
8.   Add shift s_{k+1} to shift pattern P_i with probalility p_s
9.   **Else** Add shift s_{k+1} to shift pattern P_i;
10.   **End If**;
11.   If workload w_i of sequence $[s_1 s_2 ... s_{k+1}] \geq w_{max}$ **Then**
12. $s_{k+1} = o$;
13.   **End If**;
14.   Increment counter $k = k + 1$;
15.   **Until** (Shift Pattern P_i is complete);
16.   Increment counter $i = i + 1$;
17. **Until** (Required schedules, I, are generated);
18. **Return** solution;

FIGURE 6.2   An enhanced FSM initialization procedure.

6.3.2   INITIALIZATION PHASE

The initialization algorithm generates a good initial solution, ensuring that all sequence constraints are avoided. Figure 6.2 presents an enhanced initialization algorithm, which begins by randomly generating an initial shift s_1. The algorithm then successively generates shift s_{k+1} and adds it to the existing sequence. The resulting sequence, $[s_1 s_2 ... s_{k+1}]$, is tested for membership to a set F of forbidden shifts. An example of a forbidden set is $F = \{n\text{-}d, n\text{-}e, e\text{-}d\}$. In addition, the workload of the current sequence should not exceed the maximum allowable workload w_{max}.

The initialization algorithm terminates when the required number of individual nurse schedules, I, is generated, forming a complete candidate solution representing a nurse roster.

6.3.3   GROWTH PHASE

6.3.3.1   Fuzzy Evaluation

Fuzzy evaluation measures the fitness of a candidate solution. Therefore, fitness is expressed as a weighted sum of satisfaction of soft constraints. Each soft constraint is represented as a normalized fuzzy membership function in [0,1]. Two widely accepted types of membership functions are used (Sakawa, 1993), namely, (a) triangular functions, and (b) interval-valued functions, as shown in Figure 6.3.

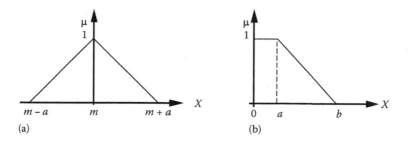

FIGURE 6.3 Linear membership functions. (a) Triangular fuzzy membership. (b) Interval-valued membership.

In Figure 6.3a, the satisfaction level is represented by a fuzzy number $A<m,a>$, where m denotes the center of the fuzzy parameter with width a. Thus, the corresponding membership function is

$$\mu_A(x) = \begin{cases} 1 - \dfrac{|m - x|}{a} & \text{If } m - a \leq x \leq m + a \\ 0 & \text{If otherwise} \end{cases} \qquad (6.1)$$

In Figure 6.3b, the satisfaction level is a decreasing linear function where $[0,a]$ is the most desirable range, and b is the maximum acceptable. The corresponding function is

$$\mu_B(x) \begin{cases} 1 & \text{If } x \leq a \\ (b - x)/(b - a) & \text{If } a \leq x \leq b \\ 0 & \text{If otherwise} \end{cases} \qquad (6.2)$$

Fuzzy multiple criteria in the nurse scheduling problem can be classified into two categories—that is, those that can be modeled by triangular fuzzy membership functions, and those that can be modeled based on interval-valued membership functions. The two sets are presented here.

Membership Function Set 1. This set assumes the triangular membership function. For example, to have a fair workload assignment, the workload h_i of each nurse i should be as close as possible to the mean workload w. This implies that workload variation $x_i = h_i - w$ should be as low as possible. Assuming a symmetrical triangular membership function, we obtain

$$\mu_1 = \mu_A(x_i) \qquad (6.3)$$

where x_i is the workload variation for nurse i from the mean w of the fuzzy parameter, with width a.

The same argument holds for the allocated days off and night shifts. A descriptive summary of the three membership functions is provided in Table 6.3.

TABLE 6.3
Description of Membership Functions, Set 1

No.	Membership	Fuzzy Parameter	Brief Description
1	μ_1	Workload	The workload of each nurse should be as close as possible to the mean workload.
2	μ_2	Days off	The allocated days off or holidays for each nurse should be as close as possible to the mean.
3	μ_3	Night shifts	The allocated number of night shifts, that is either e shift or n shift, for each nurse should be as close as possible to the mean.

Membership Function Set 2. This set assumes the interval-valued membership function. One example in this category is congeniality, which is a measure of compatibility of staff allocated similar shifts; the higher the congeniality is, the higher the schedule quality will be. In practice, a decision maker sets limits to acceptable number of uncongenial shifts x_i for each nurse i to reflect satisfaction level. Therefore, following the interval-valued functions in Figure 6.3b, the corresponding membership function is

$$\mu_4 = \mu_B(x_i) \tag{6.4}$$

where

x_i is the actual number of uncongenial allocations
a is the upper limit to the preferred uncongenial shifts
b is the maximum uncongenial shifts

As in membership functions in set 1, the same argument can be extended to number of forbidden shift sequences, number of shift variations in a single schedule, understaffing, and overstaffing. A description of the five membership functions in this category is provided in Table 6.4.

The Overall Fitness Function. For each nurse i, schedule fitness is obtained from the weighted sum of the first four membership functions. As such, the fitness for each shift pattern (or element) i is obtained according to the following expression:

$$\eta_i = \sum_{z=1}^{5} w_z \mu_z(x_i) \quad \forall i \tag{6.5}$$

where w_z is the weight of each function μ_z, such that condition $\Sigma w_z = 1.0$ is satisfied.

Similarly, the fitness according to shift requirement and congeniality in each day j is given by

$$\lambda_j = \sum_{z=6}^{7} w_z \mu_z(x_j) \quad \forall j \tag{6.6}$$

where w_z is the weight of each function μ_z, with $\Sigma w_z = 1.0$.

TABLE 6.4

Description of Membership Functions, Set 1

No.	Membership	Fuzzy Parameter	Brief Description
1	μ_4	Congeniality	Number of incompatible or uncongenial shift allocations for each nurse, where a nurse is allocated the same shift with incompatible workmates
2	μ_5	Forbidden shift sequences	Number of forbidden shift sequences should be reduced as much as possible to satisfy nurse expectations
3	μ_6	Shift variations in a schedule	Schedules with less variations (e.g., $[d\,d\,d\,o\,o]$ with a variation of 1) are more desirable than $[d\,o\,d\,o\,d]$ with a variation of 4
4	μ_7	Understaffing	Number of understaff shifts should be as low as possible; this improves service quality
5	μ_8	Overstaffing	Number of overstaffed shifts should be as low as possible; this reduces excess staffing costs

The overall fitness of the candidate solution is given by the following expression:

$$f = \left(\frac{\eta}{\omega_1} \wedge 1\right) \wedge \left(\frac{\lambda}{\omega_2} \wedge 1\right) \tag{6.7}$$

where
$\lambda = \lambda_1 \wedge \ldots \wedge \lambda_I$
$\mu = \mu_1 \wedge \mu_2 \wedge \ldots \wedge \mu_J$
I is the number of nurses
J is the number of working days
ω_1 and ω_2 are the weights associated with η and λ, respectively
"$\wedge$" is the min operator

The weights w_z, ω_1, and ω_2 enable the decision maker to incorporate his or her choices to reflect expert opinion and preferences of the management and the nurses. This feature gives FSM an added advantage over related methods.

6.3.3.2 Transformation

In NSP, elements are twofold: one that represents horizontal shift patterns, denoted by e_i, and another representing the vertical shift allocations for each day, denoted by e_j. Fitness η_i and λ_j of each element are probabilistically tested for transformation by comparing with a random number $p_t \in [0,1]$, generated at each iteration t. A dynamically decaying transformation probability limit, $p_t = p_0 e^{-t/T}$, is used.

Two procedures are incorporated into the transformation heuristics: a column-wise heuristic and a row-wise heuristic. Figure 6.4 presents the column-wise

Column-wise transformation heuristic
1.   Initialize iteration $t = 1$;
2.   **While** $(t \le t_{max})$ **do**
3.   **While** (termination condition) **do**
4.   With probability $p_c = \min[1 - \lambda, p_t]$;
5.   Randomly select c_1 = cell with conflict;
6.   Randomly select c_2 = cell with conflict, but in same column;
7.   Swap (c_1, c_2);
8.   Select the best from neighborhood;
9.   **End While**
10. $t = t + 1$;
11.   **End While**

FIGURE 6.4   Pseudocode for column-wise transformation heuristic.

Row-wise transformation heuristic
1.   Initialize iteration $t = 1$;
2.   **While** $(t \le t_{max})$ **do**
3.   **While** (termination condition) **do**
4.   With probalility $p_r = \min[1 - \lambda, p_t]$;
5.   Randomly select r_1 = cell with conflict;
6.   Randomly select r_2 = cell with conflict, but in same row;
7.   Swap (r_1, r_2);
8.   Select the best from neighborhood;
9.   **End While**
10. $t = t + 1$;
11.   **End While**

FIGURE 6.5   Pseudocode for row-wise transformation heuristic.

heuristic. The heuristic searches for improved shift sequences and schedules in the neighborhood of the current schedule for each nurse. Again, the dynamic transformation probability p_t is used to control the transformation process.

Figure 6.5 shows the row-wise heuristic. The heuristic searches for improved roster structure in the neighborhood of the current schedule for each nurse.

6.3.3.3   Regeneration

Regeneration repairs infeasible elements using a mechanism similar to the initialization algorithm, which incorporates hard constraints. Based on the juvenile hormone level m_t at iteration t, the candidate solution is then tested for readiness for maturation:

$$m_t = 1 - (\eta_1 \wedge \eta_2 \wedge \eta_3 \wedge \eta_4) \wedge (\lambda_1 \wedge \lambda_2) \tag{6.8}$$

The growth phase repeats until a predefined acceptable dissatisfaction m_0 is reached. However, the algorithm proceeds to the maturation phase if there is no significant change ε in m_t, with the value of change ε set in the order of 10^{-6}.

6.3.4 MATURATION PHASE

Intensification ensures a complete search for a near-optimal solution in the neighborhood of the current solution. In the postprocessing stage, the user interactively makes expert changes to the candidate solution, as well as to execute intensification. Expert knowledge and intuition are coded in the form of possible adjustments through weights w and ω. Illustrative computations are presented in the next section.

6.4 COMPUTATIONAL ANALYSIS

The proposed FSM algorithm was coded in Java and implemented on a 3.06 GHz speed processor, with 4GB RAM. Computational experiments and results are presented in this section.

6.4.1 COMPUTATIONAL EXPERIMENTS

To illustrate the effectiveness of the proposed FSM algorithm, computational experiments were carried out on typical nurse scheduling problems in the literature. Three sets of problem cases were used for the experiments: (1) experiment 1, a preliminary experiment adapted from Jan, Yamamoto, and Ohuchi (2000); (2) experiment 2, an extension of the problem case in experiment 1; (3) experiment 3, comprising a set of 20 benchmark problem cases in the literature (Musliu, 2006); and (4) experiment 4, consisting of extensions from the benchmark problems in (3). Problem cases in experiment 3 were obtained from real-life situations in healthcare organizations, as reported by Musliu (2006). Each experiment included constraints on shift sequences, length of shift sequences, and length of work and days off. The number of employees (or groups) for the problems ranged from 7 to 163, to be scheduled over three standard shifts—day, evening and night shifts.

The termination criteria are controlled by two conditions: (1) the maximum number of iterations, set at $T_m = 300$, and (2) the maximum number of iterations with no improvement, set at $T_I = 30$. This implies that the algorithm terminates when either of the conditions is met. Generally, each experiment was executed 50 independent times. Computational results and discussions are presented in the next section.

6.4.2 RESULTS AND DISCUSSION

6.4.2.1 Experiment 1

The first experimental problem was adapted from Jan, Yamamoto, and Ohuchi (2000). In this problem, there are 15 nurses to be scheduled over a planning horizon of 30 days. Shift sequences n-d, e-n, and e-d are unsatisfactory. The daily requirements for shifts d, e, and n are 11, 2, and 2, respectively. The day off, o, and congeniality preferences were not considered. The initial schedule with this setup is shown in Figure 6.6. The fitness values for individual nurses are very low; therefore, the schedule quality is unsatisfactory as can be seen from the low overall fitness.

Figure 6.7 shows the final optimal schedule obtained in the preliminary experiments. The overall fitness for the best solution is 1.00. This demonstrates that all the

Days

Nurse	1	2	3	4	5	6	7	8	9	10	11	12	13	14	15	16	17	18	19	20	21	22	23	24	25	26	27	28	29	30	Fitness η_i
1	e	e	e	e	e	e	e	e	e	e	e	e	e	e	e	e	e	e	e	e	e	e	e	e	e	e	e	e	e	e	0.31
2	d	d	d	e	d	d	d	d	d	d	d	d	d	d	d	d	d	d	d	d	d	d	d	d	d	d	d	d	d	d	0.33
3	d	d	d	e	d	d	d	d	d	d	d	d	d	d	d	d	d	d	d	d	d	d	d	d	d	d	d	d	d	d	0.33
4	e	e	e	e	d	d	d	d	d	d	d	d	d	d	d	d	d	d	d	d	d	d	e	e	e	e	e	e	e	e	0.54
5	d	d	d	e	d	d	d	d	d	d	d	d	d	d	d	d	d	d	d	d	d	d	d	d	d	d	d	d	d	d	0.66
6	n	n	n	n	n	n	n	n	n	n	n	n	n	n	n	n	n	n	n	n	n	n	n	n	n	n	n	n	n	n	0.49
7	d	d	d	d	d	d	d	d	d	d	d	d	d	d	d	d	d	d	d	d	d	d	d	d	d	d	d	d	d	d	0.33
8	d	d	d	d	d	d	d	d	d	d	d	d	d	d	d	d	d	d	d	d	d	d	d	d	d	d	d	d	d	d	0.50
9	n	n	n	d	d	d	d	d	d	d	d	d	d	d	d	d	d	d	d	d	d	d	d	d	d	d	d	d	d	d	0.49
10	d	d	d	d	d	d	d	d	d	d	d	d	d	d	d	d	d	d	d	d	d	d	d	d	d	d	d	d	d	d	0.50
11	n	n	n	n	n	n	n	n	n	n	n	n	n	n	n	n	n	n	n	n	n	n	n	n	n	n	n	n	n	n	0.49
12	d	d	d	d	d	d	d	d	d	d	d	d	d	d	d	d	d	d	d	d	d	d	d	d	d	d	d	d	d	d	0.50
13	d	d	d	d	d	d	d	d	d	d	d	d	d	d	d	d	d	d	d	d	d	d	d	d	d	d	d	d	d	d	0.50
14	d	d	d	e	d	d	d	d	d	d	d	d	d	d	d	d	d	d	d	d	d	d	d	d	d	d	e	e	e	e	0.57
15	e	e	e	e	e	e	e	e	e	e	e	e	e	e	e	e	e	e	e	e	e	e	e	e	e	e	e	e	e	e	0.31
Fitness λ_i	0.3	0.7	0.7	0.7	1.0	1.0	1.0	1.0	1.0	1.0	1.0	1.0	1.0	1.0	1.0	1.0	1.0	1.0	1.0	1.0	1.0	0.7	0.7	0.7	0.7	0.7	0.7	0.3	0.3	0.3	0.33

FIGURE 6.6 Initial nurse schedule for experiment 1.

Nurse	Days 1	2	3	4	5	6	7	8	9	10	11	12	13	14	15	16	17	18	19	20	21	22	23	24	25	26	27	28	29	30	Fitness η
1	e	e	d	d	d	d	d	d	d	d	d	d	d	n	n	e	e	e	d	d	d	d	d	d	d	d	d	d	n	n	1.000
2	n	e	e	d	d	d	d	d	d	d	d	d	d	d	n	n	e	e	d	d	d	d	d	d	d	d	d	d	n	n	1.000
3	n	n	e	e	d	d	d	d	d	d	d	d	d	d	d	n	n	e	e	d	d	d	d	d	d	d	d	d	d	d	1.000
4	d	n	n	e	e	d	d	d	d	d	d	d	d	d	d	d	n	n	e	e	d	d	d	d	d	d	d	d	d	d	1.000
5	d	d	n	n	e	e	d	d	d	d	d	d	d	d	d	d	d	n	n	e	e	d	d	d	d	d	d	d	d	d	1.000
6	d	d	d	n	n	e	e	d	d	d	d	d	d	d	d	d	d	d	n	n	e	e	d	d	d	d	d	d	d	d	1.000
7	d	d	d	d	n	n	e	e	d	d	d	d	d	d	d	d	d	d	d	n	n	e	e	d	d	d	d	d	d	d	1.000
8	d	d	d	d	d	n	n	e	e	d	d	d	d	d	d	d	d	d	d	d	n	n	e	e	d	d	d	d	d	d	1.000
9	d	d	d	d	d	d	n	n	e	e	d	d	d	d	d	d	d	d	d	d	d	n	n	e	e	d	d	d	d	d	1.000
10	d	d	d	d	d	d	d	n	n	e	e	d	d	d	d	d	d	d	d	d	d	d	n	n	e	e	d	d	d	d	1.000
11	d	d	d	d	d	d	d	d	n	n	e	e	d	d	d	d	d	d	d	d	d	d	d	n	n	e	e	d	d	d	1.000
12	d	d	d	d	d	d	d	d	d	n	n	e	e	d	d	d	d	d	d	d	d	d	d	d	n	n	e	e	d	d	1.000
13	d	d	d	d	d	d	d	d	d	d	n	n	e	e	d	d	d	d	d	d	d	d	d	d	d	n	n	e	e	d	1.000
14	d	d	d	d	d	d	d	d	d	d	d	n	n	e	e	d	d	d	d	d	d	d	d	d	d	d	n	n	e	e	1.000
15	e	d	d	d	d	d	d	d	d	d	d	d	n	n	e	e	d	d	d	d	d	d	d	d	d	d	d	n	n	e	1.000
Fitness λ_i	1.0	1.0	1.0	1.0	1.0	1.0	1.0	1.0	1.0	1.0	1.0	1.0	1.0	1.0	1.0	1.0	1.0	1.0	1.0	1.0	1.0	1.0	1.0	1.0	1.0	1.0	1.0	1.0	1.0	1.0	

FIGURE 6.7 Final nurse schedule for experiment 1.

TABLE 6.5
Comparative Performance Based on Experiment 1

Approach	Reference	Best Fitness	Success Rate (%)	CPU Time (sec)	Iterations
Basic CGA	Jan, Yamamoto, and Ohuchi (2000)	1.00	8.33	[a]	[a]
CGA	Jan, Yamamoto, and Ohuchi (2000)	1.00	100	49.00	100
FSM		1.00	100	32.40	40

[a] Value not provided.

desires and preferences are satisfied and the solution is desirable to patients, staff, and management, according to their expectations.

Table 6.5 compares the performance of FSM against the basic cooperative genetic algorithm (basic CGA) and improved CGA algorithms reported in Jan, Yamamoto, and Ohuchi (2000). Out of 50 independent runs, the success rate of FSM was 100%, which is comparable to 100% for CGA with 12 independent runs. In each successful run, the FSM algorithm was able to obtain the optimal solution in less than 40 iterations, compared to 100 iterations for CGA. The average computational time was 32.40 seconds. This indicates the superior computational efficiency of FSM compared to CGA.

To further demonstrate the performance of the FMS algorithm, a plot of the intermediate solutions arrived at during the algorithm execution is presented. The overall fitness value f is plotted against number of iterations t. Figure 6.8 shows a plot of the intermediate solutions during the iterative process of the algorithm. The fitness value increased from 0.33 at the initialization stage to 1.00 at the 40th iteration, which implies that the algorithm obtained the optimum solution at the 40th iteration, though the user intended the computation to run up to 300 iterations.

6.4.2.2 Experiment 2

This experimental problem is an extension of experiment 1. Here, fuzzy multi-criteria evaluation, including day off and congeniality preferences, is fully utilized

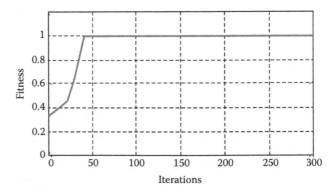

FIGURE 6.8 Illustrative computations based on problem case 1.

to determine the fitness of the candidate solution. The computational experiment consists of 15 nurses that are to be scheduled over a horizon of 30 days.

Figure 6.9 presents the initial schedule created using the enhanced initialization constructor. The daily shift requirements for shifts d, e, and n are 10, 2, and 2, respectively. Assume that, due to congeniality issues, nurse combinations (2,4) and (7,10) in any working shift are to be avoided as much as possible. The maximum number of iterations $T_m = 300$. The overall fitness at the initialization stage is 0.4667, which is very low.

Figure 6.10 shows the final nurse schedule obtained by the FSM algorithm. The solution shows a marked improvement in the fitness values of individual shift patterns. Also, there is a 100% satisfaction of the shift requirements in each day, which is a marked improvement over the initial solution. Consequently, the overall fitness value of the final schedule is 0.8197, which is a significant improvement over the initial schedule.

6.4.2.3 Further Experimental Analyses

In this experiment, computational results for 20 benchmark problems are reported. For comparative analysis, the success rate and the computational time (CPU time) are taken into consideration. For each problem, 10 independent runs were executed using the FSM algorithm. The maximum number of iterations for each run was set at $T_m = 300$.

Table 6.6 provides a summary of the comparative computational results, in terms of search success rate and average CPU time. FSM is compared with min-conflicts heuristic (MC) and MC with tabu search mechanism (MC-T), as well as the fuzzy simulated evolution (FSE) algorithm. It can be seen that FSM was able to find satisfactory solutions for all the problems—hence the 100% mean success rate, even for large-scale problems 15, 19, and 20 in the table. The success rate of FSM is comparable to MC-T, but is much better than MC and FSE. In terms of computational efficiency, FSM outperformed all the other algorithms, with a mean time of 8.17 sec, compared to 95.70 sec for MC, 20.15 for MC-T, and 9.92 for FSE. From these comparative analyses, it can be seen that FMS is capable of producing good feasible solutions satisfying patient expectations, healthcare staff preferences, and management choices.

6.4.2.4 Further Experiments

Further experiments were carried out to compare FSM and a commercial workforce scheduling software called First Class Scheduler (FCS) (Gartner, Musliu, and Slany, 2001). The comparative results of the experiments are presented in Table 6.7.

The FSM algorithm outperforms FCS, even over large problems such as 7 and 18 in the table, which have the same shift requirements over all days and shifts. The symbol "1000.0(?)" indicates that FCS could not obtain a solution for the problem within 1000 sec (Morz and Musliu, 2004; Musliu, 2006). Though FCS is known to be able to solve medium- to large-scale problems, the FSM algorithm performed better than FCS on medium- to large-scale problems. In this regard, FSM is more efficient and effective.

Days

Nurse	1	2	3	4	5	6	7	8	9	10	11	12	13	14	15	16	17	18	19	20	21	22	23	24	25	26	27	28	29	30	Fitness η_i
1	d	d	d	d	o	o	n	n	e	e	d	d	d	d	d	d		d	d	n	n	n	e	o	n	n	d	d	d	d	0.45
2	n	n	n	n	n	n	n	o	o	n	n	d	n	o	n	n	e	n	n	n	n	n	d	n	n	n	d	n	n	n	0.46
3	n	e	e	d	d	d	d	o	d	o	d	d	o	o	n	n	e	e	d	d	d	d	d	d	d	d	d	d	d	n	0.62
4	n	n	n	n	o	o	n	o	n	n	n	d	n	n	n	n	e	d	o	d	d	d	d	d	o	d	d	n	n	n	0.43
5	d	d	d	d	d	d	d	d	d	d	d	d	d	d	o	d	d	o	o	d	d	d	d	o	d	o	d	d	d	d	0.71
6	d	d	d	d	d	d	d	e	d	d	d	d	d	d	d	d	d	d	d	d	d	d	d	d	o	d	n	d	e	d	0.61
7	d	d	d	d	n	n	e	e	d	d	d	d	d	d	o	d	d	d	d	d	d	d	o	d	o	n	n	e	d	d	0.50
8	d	d	d	d	d	d	d	d	d	o	o	d	d	d	d	d	d	d	d	d	d	d	o	d	d	d	d	d	d	d	0.65
9	d	d	d	d	d	d	d	d	o	o	d	d	d	o	d	d	d	d	o	o	d	d	d	d	d	d	d	d	d	d	0.68
10	d	d	d	d	d	d	d	d	o	o	d	d	d	d	d	d	d	d	d	d	d	d	d	d	o	d	d	d	d	d	0.54
11	d	d	d	d	d	d	d	d	d	d	d	d	d	d	o	d	d	d	d	o	d	d	d	d	o	d	n	d	d	d	0.71
12	d	d	d	d	d	d	d	d	d	d	d	d	d	d	d	d	d	d	o	d	d	d	d	d	d	d	d	d	d	d	0.65
13	d	d	d	d	d	d	d	d	d	d	d	d	d	d	d	d	d	d	d	d	d	d	d	d	d	d	d	d	d	d	0.71
14	d	d	d	d	d	d	d	d	d	d	d	d	d	d	d	d	d	d	d	d	d	d	d	d	d	d	d	d	d	d	0.57
15	e	d	d	d	d	d	d	d	d	d	d	d	d	d	d	d	d	d	d	d	d	d	d	d	d	d	d	d	e	d	0.37
Fitness λ_i	0.7	0.7	0.7	0.3	0.6	0.6	0.7	0.6	0.4	0.6	0.3	0.3	0.5	0.6	0.3	0.3	0.7	0.6	0.6	0.5	0.3	0.7	0.6	0.6	0.6	0.3	0.3	0.7	1.0	0.3	0.27

FIGURE 6.9 Initial nurse schedule for experiment 2.

Nurse	1	2	3	4	5	6	7	8	9	10	11	12	13	14	15	16	17	18	19	20	21	22	23	24	25	26	27	28	29	30	Fitness η_i
1	d	d	n	n	e	e	o	d	d	d	d	d	d	d	d	o	d	d	d	d	n	n	e	e	d	d	d	d	d	d	0.77
2	d	d	d	d	d	d	d	d	d	d	d	d	n	n	n	e	d	d	d	d	d	d	d	d	d	o	n	o	n	n	0.72
3	e	d	d	d	n	d	d	d	o	d	d	n	n	e	e	e	d	d	d	d	d	d	d	d	d	n	n	n	e	e	0.70
4	d	d	n	n	e	n	e	e	d	d	d	d	d	e	e	d	d	d	e	d	d	d	d	d	o	n	n	e	e	o	0.70
5	d	n	n	e	e	d	d	d	d	d	d	o	d	n	n	n	n	e	d	d	d	d	d	n	d	d	d	d	o	d	0.82
6	e	e	o	d	d	d	d	d	d	n	n	d	d	o	o	e	e	d	e	n	e	e	n	n	e	d	o	n	n	e	0.77
7	d	d	d	d	d	d	d	d	e	d	d	e	d	d	d	d	d	d	d	e	d	d	d	n	e	d	d	d	d	d	0.75
8	o	d	d	d	d	n	d	n	n	e	e	d	d	d	o	d	d	n	n	n	e	n	e	e	d	d	d	o	o	d	0.82
9	d	d	d	d	d	d	d	e	e	d	d	d	d	d	d	d	o	d	d	d	d	d	n	d	e	d	d	d	d	d	0.79
10	n	n	e	d	d	e	d	n	d	d	e	d	d	o	o	d	n	n	e	e	o	d	n	e	d	d	d	d	d	d	0.77
11	d	o	d	n	n	d	e	o	e	o	d	d	d	d	d	d	d	e	d	n	e	e	e	e	d	n	d	e	e	e	0.79
12	d	d	d	e	o	o	n	n	e	n	n	e	d	d	e	d	d	d	o	d	n	d	e	o	d	n	n	e	d	d	0.82
13	d	d	d	d	d	d	d	d	d	o	n	n	d	d	d	d	d	d	o	d	d	o	o	d	n	n	e	e	d	d	0.78
14	d	d	d	d	d	d	d	d	n	n	e	e	d	d	d	d	d	d	d	d	d	o	o	n	n	e	e	d	d	d	0.78
15	d	d	d	d	d	d	d	d	n	n	e	e	d	d	d	d	d	d	d	d	d	d	o	o	n	n	e	d	d	d	0.76
Fitness λ_j	1.0	1.0	1.0	1.0	1.0	1.0	1.0	1.0	1.0	1.0	1.0	1.0	1.0	1.0	1.0	1.0	1.0	1.0	1.0	1.0	1.0	1.0	1.0	1.0	1.0	1.0	1.0	1.0	1.0	1.0	0.88

Days

FIGURE 6.10 Final nurse schedule for experiment 2.

TABLE 6.6
Comparison between FSM and Other Algorithms

Problem	Success Rate (%)				CPU Time (sec)			
	MC	MC-T	FSE	FSM	MC	MC-T	FSE	FSM
1	100	100	100	100	4.77	0.07	0.1	0.09
2	100	100	100	100	1.48	0.07	0.1	0.08
3	100	100	100	100	69.36	0.42	0.18	0.14
4	100	100	100	100	0.12	0.11	0.08	0.1
5	100	100	100	100	15.78	0.43	0.31	0.33
6	100	100	100	100	2.89	0.08	0.09	0.07
7	100	100	100	100	62.51	52.79	4.38	3.16
8	100	100	100	100	32.52	0.74	0.88	0.73
9	50	100	100	100	84.17	15.96	4.87	2.14
10	100	100	100	100	11.40	0.60	0.78	0.66
11	10	100	100	100	254.82	13.15	10.3	7.12
12	100	100	100	100	74.26	1.17	5.33	3.27
13	100	100	100	100	68.32	0.87	2.34	1.2
14	100	100	100	100	8.77	0.76	2.85	1.95
15	15	100	80	100	331.11	159.04	46.34	33.12
16	100	100	100	100	14.48	0.54	3.15	2.19
17	100	100	100	100	54.79	2.16	7.59	5.54
18	100	100	100	100	60.58	6.83	8.35	8.13
19	70	100	100	100	577.96	75.83	72.62	62.2
20	100	100	100	100	183.82	71.38	27.78	31.22
Mean	87.25	100.00	99.00	100.00	95.70	20.15	9.92	8.17

6.5 SUMMARY

Nurse scheduling is a common challenge in hospitals where decision makers have to assign work shifts to nurses over a planning span for a period of about a week to a month. The aim is to produce high-quality work schedules that satisfy the desires and expectations of patients, nursing staff, and management. Since these desires are uncertain in a nonstochastic sense, developing high-quality nurse schedules in such a fuzzy environment is difficult. This situation is commonplace in hospitals.

The chapter presented an enhanced fuzzy simulated metamorphosis algorithm, based on the concepts of biological evolution in insects. The algorithm mimics the hormone-controlled evolution process going through initialization, iterative growth, and maturation. The suggested methods offers a practical approach to solving fuzzy multicriteria decision problems such as nurse scheduling, task assignment, vehicle routing, and job shop scheduling. The algorithm can interactively accept users' choices, intuition, and expert opinion, which is otherwise impossible with other optimization algorithms. In addition, the algorithm uses adaptive parameters that enhance guided transformation of the solution in the search process. By using dynamic and adaptive balance between exploitation and exploration of the solution

TABLE 6.7
Further Comparison between FSM and FCS

Problem	Groups	CPU Time (sec) FCS	FSM
1	9	0.9	0.09
2	9	0.4	0.08
3	17	1.9	0.14
4	13	1.7	0.10
5	11	3.5	0.33
6	7	2.0	0.07
7	29	16.1	3.16
8	16	124.0	0.73
9	47	>1000.0(?)	2.14
10	27	9.5	0.66
11	30	367.0	7.12
12	20	>1000.0(?)	3.27
13	24	>1000.0(?)	1.20
14	13	0.54	1.95
15	64	>1000.0(?)	33.12
16	29	2.44	2.19
17	33	>1000.0	5.54
18	53	2.57	8.13
19	120	>1000.0(?)	62.2
20	163	>1000.0(?)	31.22
Mean	–	>40.97	8.17

space, the algorithm can reach the near-optimal solution in a reasonable computation time.

FSM is an invaluable addition to researchers in the operations research and operations management community, especially to those concerned with healthcare operations management. However, the application of the proposed approach can be extended to other practical problems such as task assignment, vehicle routing, job sequencing, and time tabling.

REFERENCES

Brucker, P., Burke, E. K. and Curtois, T. 2010. A shift sequence based approach for nurse scheduling and a new benchmark data set. *Journal of Heuristics* 16: 559–573.

Burke, E. K., De Causmaecker, P. and Vanden Berghe, G. 2000. Novel metaheuristic approaches to nurse rostering problems in Belgian hospitals. In *Handbook of scheduling: Algorithms, models and performance analysis*, ed. J. Leung, pp. 44.1–44.18. CRC Press, Boca Raton, FL.

Cheang, B., Li, H., Lim, A. and Rodrigues, B. 2003. Nurse rostering problems—A bibliographic survey. *European Journal of Operational Research* 151: 447–460.

Gartner, J., Musliu, N. and Slany, W. 2001. Rota: A research project on algorithms for work-force scheduling and shift design optimization. *Artificial Intelligence Communications*, 14 (2): 83–92.

Inoue, T., Furuhashi, T., Maeda, H. and Takaba, M. 2003. A proposal of combined method of evolutionary algorithm and heuristics for nurse scheduling support system. *IEEE Transactions on Industrial Electronics* 50 (5): 833–838.

Jan, A., Yamamoto, M. and Ohuchi, A. 2000. Evolutionary algorithms for nurse scheduling problem. *IEEE Proceedings of the 2000 Congress on Evolutionary Computation* 1: 196–203.

Morz, M. and Musliu, N. 2004. Genetic algorithm for rotating workforce scheduling problem. *IEEE International Conference on Computational Cybernetics* 121–126.

Musliu, N. 2006. Heuristic methods for automated rotating workforce scheduling. *International Journal of Computational Intelligence Research* 2 (4): 309–326.

Mutingi, M. and Mbohwa, C. 2013. A fuzzy genetic algorithm for healthcare staff sched-uling. *International Conference on Law, Entrepreneurship and Industrial Engineering (ICLEIE'2013)*, April 15–16, Johannesburg, South Africa, 239–243.

Mutingi, M. and Mbohwa, C. 2014a. A fuzzy-based particle swarm optimization algorithm for nurse scheduling. *IAENG International Conference on Systems Engineering and Engineering Management*, October 2014, San Francisco, 998–1003.

Mutingi, M. and Mbohwa, C. 2014b. Simulated metamorphosis—A novel optimizer. *IAENG International Conference on Systems Engineering and Engineering Management*, October 2014, San Francisco, 924–929.

Otero, F. E. B., Masegosa, A. D. and Terrazas, G. 2014. Thematic issue on advances in nature inspired cooperative strategies for optimization. *Memetic Computing* 6 (3): 147–148.

Sakawa, M. 1993. *Fuzzy sets and interactive multi-objective optimization*. Plenum Press, New York.

Shaffer, S. 1991. A rule-based expert system for automated staff scheduling. *IEEE International Conference on Systems, Man, and Cybernetics* 3: 1691–1696.

Topaloglu, S. and Selim, S. 2010. Nurse scheduling using fuzzy modeling approach. *Fuzzy Sets and Systems* 161: 1543–1563.

Vasant, P. and Barsoum, N. 2006. Fuzzy optimization of units products in mix-product selec-tion problem using FLP approach. *Soft computing. A fusion of foundations, methodolo-gies and applications* 10: 144–151.

Villacorta, P. J., Masegosa, A. D. and Lamata, M. T. 2013. Fuzzy linguistic multi-criteria morphological analysis in scenario planning. *IEEE IFSA World Congress and NAFIPS Annual Meeting (IFSA/NAFIPS)*, June 2013, 777–782.

7 Fuzzy Simulated Evolution Algorithms for Nurse Rerostering

7.1 INTRODUCTION

Healthcare systems operate in a dynamic and imprecise environment. Unanticipated events may lead to complex schedule disruptions (Moz and Pato, 2004; Knighton, 2005; Maenhout and Vanhoucke, 2011). For instance, a nurse scheduled to work in a specific shift may not be available due to unforeseen absences (Ritchie et al., 1999; Aickelin and Dowsland, 2000; Corominas and Pastor, 2010). As a result, decision makers often find it necessary to reconstruct the nurse schedules, a process known as nurse rerostering (Moz and Pato, 2007). This essentially involves reassigning shifts to available nurses, beginning from the first day of reported absence, while respecting pertinent constraints and the reported absences.

Staff scheduling and rostering problems are widely addressed in the literature. Some key literature search surveys were done by Cheang et al. (2003), Ernst et al. (2004a,b), and Burke, De Causmaek, Vanden Berghe et al. (2004). However, the rerostering problem has received very little attention. When addressing nurse scheduling problems, most researchers assume that there is always a reserve pool of nurses to replace those absent (Aickelin and Dowsland, 2000; Burke, Cowling et al., 2001; Maenhout and Vanhoucke, 2007, 2008, 2009, 2011). This may not be the case in real-world practice. Moreover, it may be costly always to have a reserve pool of nurses.

Moz and Pato (2007) presented a genetic algorithm approach to the nurse rerostering problem, based on real data from a hospital setting. In the same vein, Maenhout and Vanhoucke (2011) proposed an evolutionary metaheuristic approach for addressing the nurse rerostering problem. Pato and Moz (2008) proposed a bi-objective nurse rerostering problem based on a utopic Pareto genetic heuristic. The authors generated problem sets with artificial complexity to test the algorithm's efficiency and effectiveness. Key objectives realized in these approaches are summarized as follows:

1. To maximize or maintain the quality of service as was intended in the original roster before disruptions due to unplanned absences
2. To maximize or maintain the satisfaction of individual nurse preferences
3. To maximize or maintain schedule fairness
4. To minimize schedule changes as much as possible

In practice, all the preceding listed goals tend to be imprecise rather than crisp. The goals or targets relating to these objectives are often fuzzy (Topaloglu and Selim, 2010).

This chapter focuses on the application of the fuzzy simulated evolution algorithm to the nurse rerostering problem. The objectives are

1. To present the nurse rerostering problem
2. To apply the fuzzy simulated evolution algorithm
3. To perform illustrative computational experiments, demonstrating the effectiveness of the method

The next section presents the nurse rerostering problem, derived from extant studies in the literature (Maenhout and Vanhoucke, 2011).

7.2 THE NURSE REROSTERING PROBLEM

The nurse rerostering problem can be defined as follows: A set of n heterogeneous nurses, indexed i ($i = 1,...,n$), are scheduled over a period spanning d days, indexed j ($j = 1,...,d$). The nurses are currently assigned to one of the available shifts, indexed k ($k = 1,...,s$), where the last shift, s, is treated as the day off. In this connection, the decision for nurse rerostering is defined according to the expression

$$x_{ijk} = \begin{cases} 1 & \text{If nurse } i \text{ is scheduled to work on day } j, \text{ shift } k \\ 0 & 0 \text{ otherwise} \end{cases} \tag{7.1}$$

This implies that each available nurse is assigned to a single schedule, subject to all organizational goals, as well as labor policies. The shift assignment or roster should satisfy hard constraints affecting individual shift schedules of each nurse. In addition, decisions regarding the conflicting multiple goals are made, for instance, maximizing satisfaction of nurse preferences (Azaiez and Sharif, 2005), maximizing satisfaction of quality of patient service, minimizing understaffing and minimizing overstaffing costs, and constructing schedules that are as fair as possible.

In addressing the nurse rerostering problem, it is assumed that the nurse rostering problem has been solved satisfactorily. Following this assumption, the decision in the prior or original roster is defined as follows:

$$x'_{ijk} = \begin{cases} 1 & \text{If nurse } i \text{ was originally scheduled to work on day } j, \text{ shift } k \\ 0 & 0 \text{ otherwise} \end{cases} \tag{7.2}$$

Then, the problem of rerostering nurse schedules arises when unforeseen schedule disruptions occur due to nurse i, who can no longer work shift k on one or more of the future work days j. In this view, the rerostering problem is concerned about

	Day 1	Day 2	Day 3	Day 4
Nurse 1	D	D		N
Nurse 2	E	E	E	E
Nurse 3	N	N	N	
Nurse 4	D	D	D	D
Nurse 5			D	D
Nurse 6	D	D	D	D
ΣD	3	3	3	3
ΣE	1	1	1	1
ΣN	1	1	1	1

(a)

	Day 1	Day 2	Day 3	Day 4
Nurse 1	D		N	N
Nurse 2	E	E	E	E
Nurse 3	N	N		
Nurse 4	D	D	D	D
Nurse 5		D	D	D
Nurse 6	D	D	D	D
ΣD	3	3	3	3
ΣE	1	1	1	1
ΣN	1	1	1	1

(b)

FIGURE 7.1   A disrupted nurse schedule and a reroster. (a) Schedule disruption. (b) Suitable reroster.

TABLE 7.1
An Example of a Set of Assignable Shifts

Shift Type	Shift Description	Period
D	Day shift	8 a.m. to 4 p.m.
E	Night shift	4 p.m. to 12 a.m.
N	Late night shift	12 a.m. to 8 a.m.
O	Off day or holiday	

---

reconstructing shift schedules, based on the original schedule, over the short- to medium-term horizon.

Figure 7.1 shows an example of a nurse roster with a schedule disruption in (a) and a suitable reroster in (b). Nurses are originally assigned either day shift (D, 8 a.m. to 4 p.m.), night shift (E, 4 p.m. to 12 a.m.), or late night shift (N, 12 a.m. to 8 a.m.), shown in Table 7.1. The day off shift is represented by a blank space. Schedule disruptions are reported by nurse 1 and nurse 2 for day 2 and day 3, respectively. As in rostering, rerostering seeks to reconstruct the disrupted schedule subject to various hard and soft constraints. However, rerostering requires that schedule changes be as minimal as possible. In this view, part (b) presents a feasible roster where nurse 1 and nurse 5 are assigned the disrupted shifts on day 3 and day 2, respectively. In this case, the rerostering period spans 4 days, preferably from the day of disruption to the last day of the planning horizon.

7.2.1 PROBLEM CONSTRAINTS

There are two basic categories of nurse scheduling constraints: (1) time-related constraints, and (2) staffing requirements constraints. However, in addition to these, the nurse rerostering problem is restricted by disruption constraints.

Time-based constraints. Time-based constraints relate to labor policies, organizational regulations, and contract specifications, which control the sequence of individual nurse schedules. These are also called sequence constraints since they limit

TABLE 7.2
Other Identified Constraints

No.	Constraint	References
1	Maximum and minimum number of working days or working hours per period	Ernst et al. (2004a,b); Topaloglu and Selim (2010)
2	Number of consecutive shift assignments	Topaloglu and Selim (2010); Maenhout and Vanhoucke (2011)
3	Fair number of requested days off or holidays assigned	Inoue et al. (2003); Burke, De Causmaek, Petrovic et al. (2004); Mutingi and Mbohwa (2014)
4	Fair or equal total workload assignment	Burke et al. (2004); Mutingi and Mbohwa (2014)
	Congeniality, where two or more workmates are not compatible	Inoue et al. (2003); Mutingi and Mbohwa (2014)

the acceptable shift schedules for individual nurses. The most common time-based constraints are as follows:

1. Each nurse can be assigned at most one working shift per day; otherwise a day off is assigned.
2. A sufficient rest time should be granted where succession of certain shifts is prohibited. This refers to forbidden sequences such as night and early/late shift as well as late and early shift on the next day (Maenhout and Vanhoucke, 2011).

In addition to the common time-based constraints, other constraints are realized in the literature (Burke, De Causmaek, Petrovic et al., 2004; Maenhout and Vanhoucke, 2011; Mutingi and Mbohwa, 2014). These constraints are further identified and outlined in Table 7.2.

Staffing requirements constraints. On the other hand, staffing requirements ensure adequate coverage of healthcare tasks that need to be performed. This implies that overstaffing should be as low as possible, where zero values are most favorable:

$$\sum_k o_k \cong 0 \qquad (7.3)$$

where o_k is overstaffing for shift k.

In the same manner, understaffing should be as close to zero as possible. Therefore, this can be represented by the expression

$$\sum_k u_k \cong 0 \qquad (7.4)$$

where u_k is understaffing for shift k.

Disruption constraints. In addition to the original nurse rostering constraints, the nurse rerostering problem requires that some of the nurses not be assigned any working shift due to the reported inability to show up for duty. Therefore, due to reported unplanned absences, the following restriction is imposed as a hard constraint:

$$x_{ijk} = 0 \quad \forall (i, j, k) \in A \tag{7.5}$$

where A is a set of reported unplanned absences.

Due to the imposed disruption constraints, the roster should necessarily undergo some shift changes in order to accommodate the unplanned absences and to ensure continuity of service. However, in practice, it is essential to minimize the number of changes as much as possible in order to avoid dissatisfaction of the affected nurses. Moz and Pato (2003) revealed in their study that the primary nurse preference to be observed when rerostering is to retain the original shift assignments as much as possible.

For high-quality schedules, all the three identified types of constraints must be satisfied to the highest degree possible.

7.2.2 PROBLEM OBJECTIVES

The overall objective is to maximize the quality of a nurse roster, which includes satisfaction of patient expectations, nurse preferences, and organizational goals. Most of these decision criteria are difficult to quantify in real life. As such, the nurse rerostering problem is a multicriteria decision problem with complex imprecise or fuzzy objectives. These criteria are classified into four categories as follows:

1. Maximize or maintain quality of service
2. Maximize individual nurse preferences
3. Maximize schedule fairness
4. Minimize schedule changes

These four criteria are discussed next.

Maximize or maintain quality of service. Most decision makers in hospitals desire to ensure that a minimum level of healthcare service quality is offered. To meet this objective, a minimum number of healthcare staff is often set for each shift. On the other hand, unnecessary additional costs due to overstaffing should be avoided as much as possible. Also, additional costs due to unplanned absences, overtime, and use of standby staff should be avoided. Therefore, it follows that understaffing and overstaffing should be minimal:

$$\sum_k u_k \cong \sum_k o_k \cong 0 \tag{7.6}$$

where u_k and o_k are understaffing and overstaffing for shift k, respectively.

Maximize individual nurse preferences. High-quality schedules should satisfy individual nurse preferences (Ernst et al., 2004a,b; Bard and Purnomo, 2005). Nurse preferences define the choices of nurses to work particular shifts on specific days, including sequence-based choices. Moreover, nurses may express their wishes for the workmates with whom they prefer to work in their assigned shifts (Inoue et al., 2003). Thus, congeniality may need to be considered for high staff morale and job satisfaction (Mutingi and Mbohwa, 2014).

Maximize schedule fairness. Schedule fairness or equity is one of the most important measures of the fitness of a roster. Variables such as workload distribution can be used to measure fairness, where evenly distributed workloads imply high-quality schedules, and vice versa. This means that the schedule fairness has to be maintained when rerostering, such that all the nurses, including those reporting unplanned absences, have fair workloads. Let the workload of nurse i be denoted by ω_i. Therefore, for fairness, the variation of workload ω_i from the mean workload should be as low as possible such that

$$|\omega_i - \alpha| \cong 0 \quad \forall i \qquad (7.7)$$

where α represents the mean workload.

Minimize schedule changes. In reconstructing a roster, it is always desirable to keep the number of shift changes minimal. It follows that the new roster should be very similar to the original while taking care of all the reported disruptions. To measure similarity of the new and the old rosters, an acceptable range of the number of changes can be determined. Then, a measure of similarity can be expressed in terms of a fuzzy membership function. This will be clarified further in the fuzzy evaluation stage of the fuzzy simulated evolution algorithm presented in the next section.

7.3   FUZZY SIMULATED EVOLUTION APPROACH

Fuzzy simulated evolution (FSE) is an enhanced iterative algorithm developed from the general simulated evolution (SE) (Kling and Banejee, 1987; Ly and Mowchenko, 1993; Li and Kwan, 2001), where one or more of the original SE operators are fuzzified. FSE, like SE, is inspired by the philosophy of natural selection in biological environments. Following initialization, where a candidate solution is generated, the algorithm iteratively goes through evaluation, selection, mutation, and reconstruction operators, which work on the single candidate solution. Figure 7.2 presents the flowchart for the FSE algorithm.

In initialization, input parameters and a valid starting solution are generated. Evaluation computes the fitness of each element in the current solution. A fitness measure is used to probabilistically discard some elements in selection based on the fitness of each element. The resulting partial solution is then fed into the reconstruction operator that heuristically forms a new complete solution from the partial solution. The current complete solution is re-evaluated in a loop fashion until a termination condition is satisfied. Therefore, FSE is a search heuristic that achieves improvement through iterative perturbation and reconstruction.

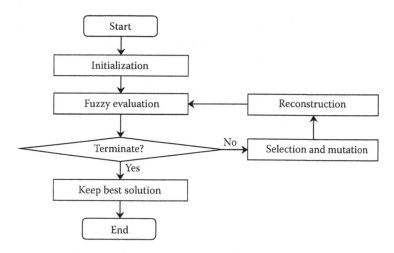

FIGURE 7.2 Fuzzy simulated evolution algorithm.

To enhance its evaluation, selection, mutation, and reconstruction processes, FSE needs to incorporate intelligent techniques such as fuzzy set theory, which enables fuzzy evaluation of candidate solutions.

7.3.1 FSE CODING SCHEME

The proposed FSE coding scheme represents a candidate solution S as a sequence of elements, where each element e_i denotes a schedule for nurse i, $i = 1,...,m$, typically covering a weekly planning horizon. This implies that each schedule is a feasible sequence of shifts D, E, N, and O for a particular nurse. A combination of schedules of all the nurses, $i = 1,...,m$, forms the overall schedule, called roster. A roster should satisfy the work requirements for each shift on each day. Furthermore, a solution space E is a set of all possible combinations of elements e_i.

Figure 7.3 shows a typical candidate solution for a complete schedule or roster. Shift "O" is represented by a blank space. The roster allocates schedules to eight nurses, covering a period of 7 days. The shift requirements for the d, e, and n shifts are 3, 2, and 2, respectively. A closer look at the proposed coding scheme reveals that evaluating a population of candidate solutions is potentially time consuming. Therefore, FSE works on a single solution to reduce computations.

7.3.2 INITIALIZATION

A good initial solution is generated as a seed for ensuing iterations. Generally, the quality of the seed influences the quality of the final solution. The FSE algorithm obtains the original roster and uses it as a seed or initial solution.

Following the initialization phase, the algorithm sequentially iterates through evaluation, selection, and reconstruction in a loop fashion till a termination criterion is satisfied. The termination criterion is defined in terms of (1) predetermined number of iterations, or (2) number of iterations without significant solution improvement.

	Day 1	Day 2	Day 3	Day 4	Day 5	Day 6	Day 7
Nurse 1	D	D		E	E	N	E
Nurse 2	E	N	E	N	E		N
Nurse 3	D	D	D	D		E	D
Nurse 4	E	N	N		D	D	D
Nurse 5	N	E	N	E	D	N	E
Nurse 6	D		D	D	N	E	D
Nurse 7	N	E	D	D	N	D	
Nurse 8		D	E	N	D	D	N
ΣD	3	3	3	3	3	3	3
ΣE	2	2	2	2	2	2	2
ΣN	2	2	2	2	2	2	2

Blank space represents shift "O"

FIGURE 7.3   Coding scheme for a typical candidate solution.

7.3.3   FITNESS EVALUATION

Fuzzy evaluation determines the fitness of the candidate solution as a function of the fitness of individual elements (nurse schedules) in the solution (roster). Thus, the aim is to determine the relative contribution of each element e_i to the fitness of the current solution S and to determine those elements that contribute below the acceptable level. The fitness of each element $F(e_i)$, is a combination of normalized functions.

The fitness or quality of a solution is a function of how much it satisfies soft constraints. As such, fitness is expressed as a function of the weighted sum of the satisfaction of the desired goals and preferences. Thus, each soft constraint is represented as a normalized fuzzy membership function in [0,1]. In this study, we use two types of membership functions: (a) triangular functions, and (b) interval-valued functions, as shown in Figure 7.4.

In Figure 7.4a, the satisfaction level is represented by a fuzzy number $A<m,a>$, where m denotes the center of the fuzzy parameter with width a. Thus, the membership function is

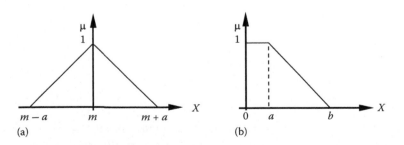

FIGURE 7.4   Linear membership functions. (a) Triangular fuzzy membership. (b) Interval-valued membership.

$$\mu_A(x) = \begin{cases} 1 - \dfrac{|m-x|}{a} & \text{If } m-a \le x \le m+a \\ 0 & \text{If otherwise} \end{cases} \qquad (7.8)$$

In Figure 7.4b, the satisfaction level is represented by a decreasing linear function where $[0,a]$ is the most desirable range, and b is the maximum acceptable. Therefore, the corresponding function is

$$\mu_B(x) = \begin{cases} 1 & \text{If } x \le a \\ (b-x)/(b-a) & \text{If } a \le x \le b \\ 0 & \text{If otherwise} \end{cases} \qquad (7.9)$$

The respective membership functions for the problem are derived, based on the previously described interval-valued functions.

Membership Function 1—Fair Workload Assignment. High-quality rosters have fair workload assignments. Therefore, the variation of workload should be as low as possible. For each nurse schedule i, the deviation x_i of workload ω_i from the average workload is

$$x_{1i} = \frac{|\omega_i - \alpha|}{\alpha} \qquad (7.10)$$

Assuming the interval-valued function, the membership function for fair workload assignment is as follows:

$$\mu_1(x_{1i}) = \mu_A(x_{1i}) \qquad (7.11)$$

where the values a and b reflect the fuzzy parameters of the interval-valued membership function.

Membership Function 2—Minimal number of shift changes. For each nurse i, let c_i be the number of shift changes, and J be the number of days or planning horizon, which is equivalent to the length of a shift pattern. It follows that satisfaction according to the objective of minimal number of changes x_{2i} for each nurse i is measured by the expression

$$x_{2i} = \frac{c_i}{J} \qquad (7.12)$$

Similarly, we assume the interval-valued membership function for fair days-off assignment. The corresponding membership function is as follows:

$$\mu_2(x_{2i}) = \mu_B(x_{2i}) \qquad (7.13)$$

where the values a and b reflect the fuzzy parameters of the membership function.

Membership Function 3—Variation of Night Shifts. For shift fairness the variation x_i of the number of night shifts (shifts e and n) allocated to each nurse i should be as close as possible to the mean allocation m. Assuming a symmetrical triangular membership function, we obtain

$$\mu_{3i}(x_{3i}) = \mu_A(x_{3i}) \tag{7.14}$$

where x_{3i} is the variation of number of night shifts allocated to nurse i from mean m of the fuzzy parameter, with width a.

Membership Function 4—Forbidden Shift Sequences. The number of shifts in the forbidden set affects the quality of the schedule for each nurse. Let the number of forbidden sequences for each nurse i be x_{4i}. The goal is to reduce the forbidden shifts as much as possible. Therefore, this can be represented by a linear interval-valued membership function as follows:

$$\mu_{4i}(x_{4i}) = \mu_B(x_{4i}) \tag{7.15}$$

where x_{4i} is the actual number of forbidden shift sequences and a and b are the fuzzy parameters of the function.

Membership Function 5—Shift Variation. For each nurse i, a schedule with a continuous sequence or block of similar shifts is often more desirable than schedules with different types of shifts. For instance, shift $[d\ d\ d\ o\ o]$ with shift variation $x_{5i} = 1$ is more desirable than shift $[d\ o\ d\ o\ d]$ with a variation $x_{5i} = 4$. Therefore, the situation can be represented by a linear interval-valued membership function:

$$\mu_{5i}(x_{5i}) = \mu_B(x_{5i}) \tag{7.16}$$

where x_{5i} is the actual number of shift variations and a and b are the fuzzy parameters of the function.

Membership Function 6—Congeniality. This membership function measures the compatibility (congeniality) of staff allocated similar shifts; the higher the congenialities are, the higher is the schedule quality. In practice, a decision maker sets limits to an acceptable number of uncongenial shifts x_{6i} for each nurse i to reflect satisfaction level. Assuming the interval-valued function, the corresponding membership function is

$$\mu_{6i} = \mu_B(x_{6j}) \tag{7.17}$$

where x_{6i} is the actual number of uncongenial allocations and a and b are the fuzzy parameters of the membership function.

Membership Function 7—Overstaffing. High-quality schedules simultaneously minimize overstaffing for each shift k. Overall, the level of overstaffing $x_{7j} = \Sigma o_k$ for all shifts in each day j should be close to zero. This can be represented by the interval-valued membership function. Therefore,

$$\mu_{7j} = \mu_B(x_{7j}) \tag{7.18}$$

where x_{7j} is the total overstaffing on day j and a and b are the fuzzy parameters of the fuzzy membership function.

Membership Function 8—Understaffing. High-quality schedules simultaneously minimize overstaffing for each shift k. Overall, the level of overstaffing $x_{8j} = \Sigma o_k$ for all shifts in each day j should be close to zero. This can be represented by a symmetrical triangular membership function. Therefore,

$$\mu_{8j} = \mu_B(x_{8j}) \qquad (7.19)$$

where x_{8j} is the total overstaffing on day j and a and m are the fuzzy parameters of the membership function.

The Overall Fitness Function. For each nurse i, schedule fitness is obtained from the weighted sum of the first four membership functions. As such, the fitness for each shift pattern (or element) i is obtained according to the following expression:

$$\eta_i = \sum_{z=1}^{6} w_z \mu_z(x_i) \quad \forall i \qquad (7.20)$$

where w_z is the weight of each function μ_z, such that condition $\Sigma w_z = 1.0$ is satisfied. Similarly, the fitness according to shift requirement in each day j is given by

$$\lambda_j = \sum_{z=7}^{8} w_z \mu_z(x_j) \quad \forall j \qquad (7.21)$$

where w_j is the weight of each function μ_j, with $\Sigma w_z = 1.0$.

From the preceding membership functions, the overall fitness of the candidate solution is given by the expression

$$f = \left(\frac{\eta}{\omega_1} \wedge 1 \right) \wedge \left(\frac{\lambda}{\omega_2} \wedge 1 \right) \qquad (7.22)$$

where
$\lambda = \lambda_1 \wedge \ldots \wedge \lambda_I$
$\mu = \mu_1 \wedge \mu_2 \wedge \ldots \wedge \mu_J$
I is the number of nurses
J is the number of working days
ω_1 and ω_2 are the weights associated with η and λ, respectively
"$\wedge$" is the min operator

The weights w_z, w_j, ω_1, and ω_2 offer the decision maker an opportunity to incorporate his or her choices reflecting expert opinion and preferences of management and nurses.

Selection algorithm	
1.	Set constant p = 0.2;
2.	Initialize i = 1;
3.	**While** $(i \leq m)$ **do**
4.	Compute fitness of element i; F_i;
5.	Let p_t = Random [0,1];
6.	Let f_t = max [0, p_t - p];
7.	**If** $(F_i < f_t)$ **Then,**
8.	discard i, return;
9.	**Else** return i;
10.	**End If;**
11.	$i = i + 1$;
12.	**End While;**
13.	**Return** Solution;

FIGURE 7.5   Algorithm for the selection phase.

7.3.4   SELECTION

The selection operator probabilistically determines whether or not an element or shift pattern I should be retained for the next generation. Elements with a high fitness value F_i have a higher probability of surviving into the next generation. Discarded elements are reserved in queue for the reconstruction phase. Selection compares fitness F_i with an allowable fitness f_t at iteration t:

$$f_t = \max [0, p_t - p] \tag{7.23}$$

where p_t is a random number in [0,1] at iteration t and p is a predetermined constant in [0,1].

Figure 7.5 presents a summary of the selection algorithm. The algorithm begins by computing the allowable fitness f_t. At each iteration t, compare fitness $F(e_i)$ of element e_i. Compare $F(e_i)$ with the allowable fitness f_t and return the element with better fitness.

The expression $f_t = p_t - p$ enhances convergence; when p_t is high, the probability of discarding good elements is very high, which is inefficient. As a result, the search power can be controlled by setting the value of p to a reasonable value (e.g., $p = 0.22$ in this study).

7.3.5   MUTATION

Mutation performs *intensive* and *exploratory* search around solution S and in unvisited regions of the solution space, respectively. Intensification is performed by swapping randomly chosen pairs of elements within a group. On the other hand, exploration enables the algorithm to move from local optima. This involves probabilistic elimination of some elements, even the best performing ones. Generally, mutation is applied at a very low probability, p_m, to ensure convergence. In this application, we use a decay function to model a dynamic mutation probability as follows:

$$p_m(t) = p_0 e^{(-t/T) \cdot \ln(2)} \tag{7.24}$$

where
   t is the iteration count
   T is the maximum count
   p_0 is the initial mutation probability

This expression can be used for both exploratory and intensive mutation probabilities. Any infeasible partial solutions are repaired in the reconstruction phase.

7.3.6 RECONSTRUCTION

The reconstruction phase rebuilds the partial solution evolved from the previous phases into a complete solution. This essentially means assigning clients to empty spaces in every incomplete group. A greed-based constructive heuristic is used for the reconstruction process, based on the attractiveness of adding a shift k into the current incomplete solution, thereby increasing the fitness F_i of a shift sequence i in that solution. The algorithm keeps a limited number of discarded elements in a set Q. Figure 7.6 shows the generalized reconstruction algorithm procedure.

Reconstruction algorithm
1.   Input incomplete solution;
2.   **For** i = 1 to I
3.   Initialize shift sequence position k = 1;
4.   **Repeat**
5.   If sequence $[s_k s_{k+1}]$ $\notin$ Forbidden set F, **Then**
6.   Insert shift s_{k+1} = rand (D, E, N);
7.   If workload w_i of sequence $[s_1 s_2 ... s_{k+1}]$ $\geq w_{max}$ **Then**
8. s_{k+1} = 0;
9.   **End If;**
10.   Increment counter $k = k + 1$;
11.   **End If;**
12.   **Until** (Shift sequence P_i is complete);
13.   Increment counter $i = i + 1$;
14.   **End For** (Required schedules, I, are generated);
15.   *//Check for shift requirements*
16.   **For** each shift k in day j
17.   If shift requirement r_k is not met, **Then**
18.   Adjust number of k shifts in the day, accordingly;
19.   **End If;**
20.   **Return** solution S;

FIGURE 7.6 FSE reconstruction algorithm.

Each shift assignment is subject to sequence and workload restrictions, where a shift "O" is assigned in the case of violation of the restrictions. The iterative loops run till each nurse is assigned a feasible shift pattern, which makes a complete roster for the nursing staff.

Subsequently, the complete roster is checked for compliance with shift requirements. This implies that the total assignment for each shift k is checked against the predetermined shift requirement r_k. In the case that requirement r_k is not met, eliminate surplus or add missing shift k accordingly. This operation is performed over all shifts in each day.

The resulting overall structure of the FSE approach is presented in the next section.

7.4   OVERALL FSEA STRUCTURE

The overall structure of the FSEA pseudocode consists of the algorithms described in the previous sections, including initialization, evaluation, selection, mutation, and reconstruction. The algorithm initially obtains the basic input from the user—that is, parameters p, p_m, p_0, and T_{max}. This is followed by generation of the initial solution S. In the evaluation phase, the fitness of each element is computed using the fuzzy evaluation method. Fitness is used to select elements that will be discarded. The resulting partial solution is then reconstructed to obtain a complete feasible solution structure. Finally, the procedure tests for termination by comparing the current iteration t with the maximum allowable T_{max}. Figure 7.7 presents a summary of the overall pseudocode for the FSEA structure.

The next sections provide numerical illustrations, results, and discussions.

7.5   ILLUSTRATIVE EXPERIMENTS

To test the efficiency and effectiveness of the FSE algorithm, complex data sets were obtained from literature (Mutingi and Mbohwa, 2014), and some were artificially generated. The first test data presented here assume that there are no days off, and a perfect initial roster satisfying all preference constraints is disrupted by reported absences from nurses 1, 5, 8, and 12, as shown in Figure 7.8. The nurses report that

The overall FSE structure	
1.	Input; p, p_m, p_0, T_{max};
2.	Initialize; randomly generate solution S;
3.	**Repeat**
4.	Evaluation;
5.	Selection(Discard);
6.	Mutation;
7.	Reconstruction;
8.	Termination condition, $t = t + 1$;
9.	**Until** $(t \geq T_{max})$;
10.	Return best solution S^*;

FIGURE 7.7   The overall FSEA pseudocode.

	1	2	3	4	5	6	7	8	9	10	11	12	13	14	15	16	17	18	19	20	21	22	23	24	25	26	27	28	29	30	Fitness η_i
Nurse 1	E	E	E	D	D	D	D	D	D	D	D	D	N	N	E	E	E	D	D	D	D	D	D	D	D	D	D	D	N	N	1.000
Nurse 2	N	E	E	E	D	D	D	D	D	D	D	D	D	N	E	E	E	D	D	D	D	D	D	D	D	D	D	D	D	N	1.000
Nurse 3	N	N	E	E	E	D	D	D	D	D	D	D	D	N	N	E	E	E	D	D	D	D	D	D	D	D	D	D	D	D	1.000
Nurse 4	D	N	N	E	E	E	D	D	D	D	D	D	D	D	N	N	E	E	E	D	D	D	D	D	D	D	D	D	D	D	1.000
Nurse 5	D	D	N	N	E	E	D	D	D	D	D	D	D	D	D	N	N	E	E	E	D	D	D	D	D	D	D	D	D	D	1.000
Nurse 6	D	D	D	N	N	E	E	D	D	D	D	D	D	D	D	D	N	N	E	E	E	D	D	D	D	D	D	D	D	D	1.000
Nurse 7	D	D	D	D	N	N	E	E	D	D	D	D	D	D	D	D	D	N	N	E	E	E	D	D	D	D	D	D	D	D	1.000
Nurse 8	D	D	D	D	D	D	N	N	E	E	D	D	D	D	D	D	D	D	N	N	E	E	E	D	D	D	D	D	D	D	1.000
Nurse 9	D	D	D	D	D	D	D	N	N	E	E	D	D	D	D	D	D	D	D	N	N	E	E	E	D	D	D	D	D	D	1.000
Nurse 10	D	D	D	D	D	D	D	D	N	N	E	E	D	D	D	D	D	D	D	D	N	N	E	E	E	D	D	D	D	D	1.000
Nurse 11	D	D	D	D	D	D	D	D	D	N	N	E	E	D	D	D	D	D	D	D	D	N	N	E	E	E	D	D	D	D	1.000
Nurse 12	D	D	D	D	D	D	D	D	D	N	N	E	E	D	D	D	D	D	D	D	D	N	N	E	E	E	D	D	D	D	1.000
Nurse 13	D	D	D	D	D	D	D	D	D	D	N	N	E	E	D	D	D	D	D	D	D	D	N	N	E	E	E	D	D	D	1.000
Nurse 14	E	D	D	D	D	D	D	D	D	D	D	N	N	E	E	D	D	D	D	D	D	D	D	N	N	E	E	E	D	D	1.000
Nurse 15	E	D	D	D	D	D	D	D	D	D	D	D	N	N	E	E	D	D	D	D	D	D	D	D	N	N	E	E	N	E	1.000
Fitness λ_j	1.0	1.0	1.0	1.0	1.0	1.0	1.0	1.0	1.0	1.0	1.0	1.0	1.0	1.0	1.0	1.0	1.0	1.0	1.0	1.0	1.0	1.0	1.0	1.0	1.0	1.0	1.0	1.0	1.0	1.0	

FIGURE 7.8 Initial roster with disruptions as indicated.

	1	2	3	4	5	6	7	8	9	10	11	12	13	14	15	16	17	18	19	20	21	22	23	24	25	26	27	28	29	30	Fitness η_i
Nurse 1	D	D	N	E	D	D	D	D	N	N	E	D	D	D	N	N	E	E	D	D	D	D	D	N	N	E	E	D	D	D	1.000
Nurse 2	N	E	E	N	E	D	D	D	D	D	D	D	D	D	D	D	D	D	D	D	D	D	D	D	D	D	D	D	D	N	1.000
Nurse 3	N	N	E	E	D	D	D	D	D	N	N	E	E	D	D	D	D	N	N	E	E	D	D	D	N	N	E	E	D	N	1.000
Nurse 4	D	N	N	E	E	D	D	D	D	D	D	D	D	D	D	D	N	N	E	E	D	D	D	E	D	D	D	D	D	D	1.000
Nurse 5	D	D	D	N	N	E	E	D	D	D	D	D	D	E	D	D	D	D	N	N	E	E	D	D	D	D	D	D	D	D	1.000
Nurse 6	D	D	D	N	N	E	E	D	D	D	D	D	D	D	D	D	D	D	N	N	E	E	D	D	D	D	D	D	D	D	1.000
Nurse 7	D	D	D	D	N	N	E	E	D	D	D	D	D	D	D	D	D	D	D	N	N	E	E	D	D	D	D	D	D	D	1.000
Nurse 8	D	D	D	N	N	E	E	D	D	D	D	D	D	D	D	D	D	D	D	N	E	E	D	D	D	D	D	D	D	D	1.000
Nurse 9	D	D	D	D	D	D	D	N	N	E	E	D	D	D	D	D	D	D	D	D	D	D	N	N	E	E	D	D	D	D	1.000
Nurse 10	D	D	D	D	D	D	D	D	N	N	E	E	D	D	D	D	D	D	D	D	D	D	D	N	N	E	E	D	D	D	1.000
Nurse 11	E	E	D	D	D	D	D	N	N	E	E	D	D	N	E	D	D	D	D	D	D	D	D	D	D	D	D	N	N	E	1.000
Nurse 12	E	D	D	D	D	D	D	D	D	D	D	D	N	N	E	E	D	D	D	D	D	D	D	D	D	N	N	E	E	D	1.000
Nurse 13	D	D	D	D	D	D	D	D	D	D	D	D	N	N	E	E	D	D	D	D	D	D	D	D	D	N	N	E	E	D	1.000
Nurse 14	D	D	D	D	D	D	D	D	D	D	D	D	N	N	E	E	D	D	D	D	D	D	D	D	D	N	N	E	E	D	1.000
Nurse 15	D	D	D	D	D	D	D	D	D	D	D	D	N	N	E	E	D	D	D	D	D	D	D	D	D	N	N	E	E	D	1.000
Fitness λ_d	1.0	1.0	1.0	1.0	1.0	1.0	1.0	1.0	1.0	1.0	1.0	1.0	1.0	1.0	1.0	1.0	1.0	1.0	1.0	1.0	1.0	1.0	1.0	1.0	1.0	1.0	1.0	1.0	1.0	1.0	

FIGURE 7.9   Final re-scheduled roster.

they can only show up for shifts other than the ones indicated in the shaded areas. The aim is to reconstruct the roster so that the disruption constraints are satisfied, while minimizing the total number of changes to the original roster.

The second experiment seeks to show the effectiveness and efficiency of the algorithm over larger scale problems. As such, problems with increasing sizes were artificially created to test the algorithm. Computational results and discussions are presented in the next section.

7.6 RESULTS AND DISCUSSIONS

Figure 7.9 shows the computational results for the first experiment. Due to unplanned absences of nurses 1, 5, 8, and 12, the roster was rescheduled, yet with minimal changes to the original roster. It is interesting to note that the overall satisfaction of the new roster is still at an acceptable level of 1.00. This demonstrates that the FSE algorithm can satisfactorily address complex multicriteria rerostering problems even in the presence of fuzzy goals and preference constraints. The algorithm has potential to solve large-scale problems with reasonable computation time.

Table 7.3 shows the performance of FSE for problems with varying sizes. For each problem size varying from 7 nurses to 25 nurses, the total reported absences varied from three to nine. All the solutions were run until the new roster was within 80% to 100% similarity to the original roster.

It can be observed from the results that the FSE algorithm is efficient and effective. All solutions obtained were satisfactory, above 85% similarity to the original roster. The maximum computation time was 304.6 seconds, which is reasonably low. Furthermore, the variation of computation time with problem size was not significant, revealing that the algorithm can be effective even over large-scale problem sizes.

TABLE 7.3
Computation[a] for Problems with Varying Sizes

		Absences						
Problem	Nurses	3	4	5	6	7	8	9
P1	7	12.2	11.2	21.1	28.7	31.2	44.8	67.8
P2	10	12.8	18.6	32.0	38.6	44.8	62.9	73.1
P3	12	26.1	26.6	33.7	55.3	56.7	69.0	98.4
P4	14	26.8	32.7	56.3	80.2	81.6	82.1	111.2
P5	15	33.5	49.8	77.3	88.0	97.8	137.5	136.6
P6	18	36.0	77.0	91.6	93.1	111.3	148.1	188.4
P7	20	48.7	110.1	107.9	109.7	188.4	194.9	235.7
P8	25	74.2	121.7	128.0	193.3	201.2	286.3	304.6

[a] In seconds.

7.7 SUMMARY

Nurse rerostering is a multicriteria decision problem aimed at satisfying rebuilding original nurse rosters to the satisfaction of (1) patient expectations on healthcare service quality, (2) staff preferences over shift sequences and workload assignments, and (3) management goals. Since human expectations, preferences, and expectations are often fuzzy, constructing high-quality schedules is inherently difficult, yet crucial. As such, the nurse rerostering problem is a concern in most hospitals. Schedule quality potentially improves worker morale and avoids absenteeism and attrition. In an environment where human preferences and expectations are imprecise, the use of fuzzy set theory concepts is beneficial. This chapter proposed an FSE algorithm that incorporates a fuzzy multicriteria fitness evaluation method, with heuristic perturbation and improvement heuristics.

Experimental results demonstrated that FSE is capable of solving medium- to large-scale nurse scheduling problems within a reasonable computation time. The algorithm was able to solve medium- to large-scale problems, with average computational time of the algorithm ranging from a few seconds to less than 304 seconds. This showed that FSEA is computationally efficient and effective.

By providing the user an opportunity to use weights, the decision maker can incorporate preferences and choices in an interactive manner. The FSEA algorithm allows interactive decision making that provides a list of good alternative solutions. This is more acceptable to practicing decision makers than prescriptive optimization methods that provide a single solution. Hence, the decision maker can use expert opinion deriving from information from patients, nurses, and managers to make adjustments to the solution process based on weights. Therefore, FSE is an effective and efficient approach for decision support in nurse rerostering.

REFERENCES

Aickelin, U. and Dowsland, K. 2000. Exploiting problem structure in a genetic algorithm approach to a nurse rostering problem. *Journal of Scheduling* 3: 139–153.

Azaiez, M. and Al Sharif, S. 2005. A 0–1 goal programming model for nurse scheduling. *Computers & Operations Research* 32: 491–507.

Bard, J. and Purnomo, H. 2005. Preference scheduling for nurses using column generation. *European Journal of Operational Research* 164: 510–534.

Burke, E., Cowling, P., De Causmaecker, P. and Vanden Berghe, G. 2001. A memetic approach to the nurse rostering problem. *Applied Intelligence* 15: 192–214.

Burke, E., De Causmaecker, P., Petrovic, S. and Vanden Berghe, G. 2001. Fitness evaluation for nurse scheduling problems. *Proceedings of Congress on Evolutionary Computation*, CEC2001: 1139–1146.

Burke, E., De Causmaeker, P., Petrovic, S. and Vanden Berghe, G. 2004. Variable neighborhood search for nurse rostering problems. In *Metaheuristics: Computer decision-making*, ed. M. Resende and S. J. Pinhode, 153–172. Kluwer Academic Publishers, Boston.

Burke, E., De Causmaecker, P., Vanden Berghe, G. and Van Landeghem, H. 2004. The state of the art of nurse rostering. *Journal of Scheduling* 7: 441–499.

Cheang, B., Li, H., Lim, A. and Rodrigues, B. 2003. Nurse rostering problems—A bibliographic survey. *European Journal of Operational Research* 151: 447–460.

Corominas, A. and Pastor, R. 2010. Replanning working time under annualized working hours. *International Journal of Production Research* 48: 1493–1515.

Ernst, A., Jiang, H., Krishnamoorthy, M., Owens, B. and Sier, D. 2004a. An annotated bibliography of personnel scheduling and rostering. *Annals of Operations Research* 127: 21–144.

Ernst, A., Jiang, H., Krishnamoorthy, M. and Sier, D. 2004b. Staff scheduling and rostering: A review of applications, methods and models. *European Journal of Operational Research* 153: 3–27.

Inoue, T., Furuhashi, T., Maeda, H. and Takaba, M. 2003. A proposal of combined methods of evolutionary algorithm and heuristics for nurse scheduling support system. *IEEE Transactions on Industrial Electronics* 50: 833–838.

Kling, R.M. and Banejee, P. 1987. ESP: A new standard cell placement package using simulated evolution. *Proceedings of the 24th ACWIEEE Design Automation Conference*, pp. 60–66.

Knighton, S. 2005. An optimal network-based approach to scheduling and re-rostering continuous heterogeneous workforces. PhD thesis, Arizona State University, Tempe.

Li, J. and Kwan, R.S.K. 2001. A fuzzy simulated evolution algorithm for the driver scheduling problem. *Proceedings of the 2001 IEEE Congress on Evolutionary Computation, IEEE Service Center*, pp. 1115–1122.

Ly, T.A. and Mowchenko, J.T. 1993. Applying simulated evolution to high level synthesis. *IEEE Transaction on Computer-Aided Design of Integrated Circuits and Systems* 12: 389–409.

Maenhout, B. and Vanhoucke, M. 2007. An electromagnetic meta-heuristic for the nurse scheduling problem. *Journal of Heuristics* 13: 359–385.

Maenhout, B. and Vanhouke, M. 2008. Comparison and hybridization of crossover operators for the nurse scheduling problem. *Annals of Operations Research* 159: 333–353.

Maenhout, B. and Vanhouke, M. 2009. Branching strategies in a branch-and-price approach for a multiple objective nurse scheduling problem. *Journal of Scheduling* 13: 77–93.

Maenhout, B. and Vanhouke, M. 2011. An evolutionary approach for the nurse rerostering problem. *Computers and & Operations Research* 1400–1411.

Moz, M. and Pato, M. 2003. An integer multicommodity flow model applied to the rerostering of nurse schedules. *Annals of Operations Research* 119: 285–301.

Moz, M. and Pato, M. 2004. Solving the problem of rerostering nurse schedules with hard constraints: New multi-commodity flow models. *Annals of Operations Research* 128: 179–197.

Moz, M. and Pato, M. 2007. A genetic algorithm approach to a nurse rerostering problem. *Computers and Operations Research* 34: 667–691.

Mutingi, M. and Mbohwa, C. 2014. A fuzzy-based particle swarm optimization algorithm for nurse scheduling. *IAENG International Conference on Systems Engineering and Engineering Management*, October 2014, San Francisco, 998–1003.

Pato, M. and Moz, M. 2008. Solving a bi-objective nurse rerostering problem by using a utopic Pareto genetic heuristic. *Journal of Heuristics* 14: 359–374.

Ritchie, K., Macdonald, E., Gilmour, W. and Murray, K. 1999. Analysis of sickness absence among employees of four NHS trusts. *Occupational Environmental Medicine* 56: 702–708.

Topaloglu, S. and Selim, S. 2010. Nurse scheduling using fuzzy modeling approach. *Fuzzy Sets and Systems* 161: 1543–1563.

8 Fuzzy Particle Swarm Optimization for Physician Scheduling

8.1 INTRODUCTION

In most healthcare organizations, constructing high-quality and fair work schedules is a difficult but critical task for decision makers. High-quality work schedules should satisfy patient expectations and personnel preferences, as well as organizational goals and policies (Burke et al., 2006; Lin and Yeh, 2007; Topaloglu and Selim, 2010). In turn, this will have important implications on quality of service, staff morale, resource utilization, and overall organizational performance (Bard and Purnomo, 2005; Topaloglu and Selim, 2007). Consequently, much research attention has been focused on staff scheduling or staff rostering in several industrial disciplines (Ernst et al., 2004a,b), especially in healthcare organizations (Cheang et al., 2003; Mutingi and Mbohwa, 2013).

Staff scheduling in healthcare organizations can be classified into three basic categories: (1) nurse rostering (Cheang et al., 2003), (2) homecare staff scheduling (Mutingi and Mbohwa, 2015), and (3) physician scheduling (Puente et al., 2009). While nurse rostering is concerned with constructing work shifts for nurses in a hospital ward setting, homecare staff scheduling involves spatial factors where healthcare staff have to visit and assist or treat patients in their homes. On the other hand, physician scheduling seeks to construct work schedules for 24-hour specialist-led emergency care in a medical treatment facility (Lo and Lin, 2011).

Physician scheduling in a hospital emergency department (HED) is comparable to nurse rostering. For example, both problems seek to find work shift assignments that satisfy the desired service goals, the patient expectations, the preferences of the concerned healthcare staff, and the organizational goals and policies. These desires and expectations complicate the nature of the two problems. As a result, constructing work schedules becomes too difficult and time consuming. However, physician scheduling has more complex characteristics than nurse rostering. For instance, the variability and uncertainty of patient demand is high, the range of the demanded services is high, and the nature of work involved is different (e.g., emergency care with other departments, outpatient service, mobile intensive care units, and general emergencies) (Puente et al., 2009). Moreover, the labor contracts in physician scheduling may differ from those in nurse rostering, in terms of skills and full-time or part-time status. Table 8.1 presents a comparison of nurse rostering and physician scheduling in a hospital emergency department.

TABLE 8.1

A Comparison of the Nurse Rostering and Physician Scheduling

Characteristics	Nurse Rostering	Physician Scheduling
Demand uncertainty	O O O	O O O O O
Work complexity	O O	O O O O
Labor mix	O O	O O O
Complexity of goals	O O O	O O O O
Complexity of constraints	O O O O	O O O O O
Planning horizon	O O	O O

Scarce research attention has been given to solving the physician scheduling problem. The purpose of this chapter is to develop a fuzzy particle swarm optimization approach to physician scheduling in an HED. The milestones in this research are as follows:

1. To investigate the related extant approaches to staff scheduling problems
2. To develop a fuzzy particle swarm optimization algorithm for physician scheduling
3. To apply the algorithm to sample problems, showing its effectiveness and efficiency

In the next section, related extant work is reviewed, highlighting the staff scheduling problems addressed, the methods applied, and the weaknesses observed.

8.2 RELATED WORK

When addressing staff scheduling problems, in general, and physician scheduling problems in particular, a number of possible approaches may need to be considered. Primarily, mathematical programming (MP) methods, such as linear programming, goal programming, and multiobjective MP, can be utilized (Bailey, 1985; Beaumont, 1997; Ernst et al., 2004a,b). However, these methods are not suitable for large-scale problems with multiple conflicting goals. Moreover, they cannot handle complex problems satisfactorily. For instance, the methods may not be applicable in the case of imprecise or fuzzy employee preferences and management goals (Topaloglu and Selim, 2010).

Recent alternative methods utilize constraint-based programming and, most importantly, heuristic and metaheuristic algorithms. Several of these approaches exist in the literature (Cheang et al., 2003; Ernst et al., 2004a,b; Mutingi and Mbohwa, 2013). Cheang et al. (2003) identified two broad categories of solution approaches to nurse rostering—namely, optimization approaches and decision approaches. Optimization approaches are usually based on MP techniques, while decision approaches usually employ heuristics and other artificial intelligences tools.

On the other hand, Ernst et al. (2004a,b) identified 29 specific categories of staff scheduling approaches, including, simple local search, artificial intelligence, expert systems, genetic algorithms (GAs), constructive heuristics, constraint logic programming, simulated annealing (SA), integer programming, set partitioning, queuing theory, and simulation.

Recently, Bergh et al. (2013) classified staff scheduling solution approaches into seven categories: mathematical programming, constructive heuristics, improvement heuristics, simulation, constraint programming, queuing theory, and others. It is important to note that though these methods are able to meet the basic requirements of the scheduling problems, the following voids are observed:

1. Most practical scheduling problems have vague goals and constraints that cannot be represented with any degree of real precision in the search algorithm.
2. Real-world staff scheduling problems have multiple decision criteria where decision makers have to seek a cautious trade-off between the decision criteria.
3. The solution space is usually too extensive for basic search methods to work efficiently and effectively.

Despite these drawbacks, a number of heuristic applications exist in the literature. Most of these applications focus on nurse scheduling (Cheang et al., 2003; Ernst et al., 2004a,b). For instance, genetic algorithms (Aickelin and Dowsland, 2004; Puente et al., 2009), simulated annealing (Bailey, Garner, and Hobbs, 1997), tabu search (Dowsland, 1998; Bester, Nieuwoudt, and Van Vuuren, 2007), hyperheuristics (Burke, Kendall, and Soubeiga, 2003), and memetic evolutionary algorithms (Aickelin et al., 2007) have been applied in a number of problems. Other applications in the healthcare sector are homecare nurse scheduling (Mutingi and Mbohwa, 2015), patient assignment (Ferrin et al., 2004; Hertz and Lahrichi, 2009), care task assignment (Cheng et al., 2007), bed allocation (Kim et al., 2000; Gorunescu, McClean, and Millard, 2002; Gong, Zhang, and Fan, 2010), patient scheduling (Daknou et al., 2010), and physician scheduling (Puente et al., 2009; Lo and Lin, 2011). However, among all these healthcare operations problems, the physician scheduling problem has received little attention in the operations management community. The research focus of this chapter is centered on the physician scheduling problem, with application of a fuzzy-based particle swarm optimization algorithm.

The next section presents a description of the physician scheduling problem, from a multicriteria perspective.

8.3　THE PHYSICIAN SCHEDULING PROBLEM

The physician scheduling problem is concerned with the efficient and fair assignment of different shift types to a given set of physicians. For instance, day shift (8 a.m. to 3 p.m.), evening shift (3 p.m. to 10 p.m.), and night shift (10 p.m. to 8 a.m. of the next day) are possible shift types. In addition, standby shifts such as weekday

TABLE 8.2

Examples of Shift Types in Physician Scheduling

Shift Type	Shift Description	Time Slot
D	Morning shift	0800 to 1500 hours
E	Afternoon shift	1500 to 2200 hours
N	Night shift	2200 to 0800 hours
W	Standby on a weekday	0800 to 0800 hours
H	Standby on a holiday	1000 to 1000 hours
Blank	Day off	–

and holiday standby shifts (8 a.m. to 8 a.m. of the next day) are also possible. Table 8.2 lists and describes typical shift types for physician scheduling.

Shift assignment is always restricted by the existing labor laws and contractual requirements of the local hospital (Ernst et al., 2004a,b; Puente et al., 2009; Lo and Lin, 2011). In the case of a mixed workforce with permanent and temporary or part-time physicians, assigning permanent staff is often most favorable. As a result, the physician scheduling problem is highly constrained. Hard constraints must be satisfied, while soft constraints may be violated at a cost.

Oftentimes, hard constraints involve demand coverage constraints, while soft constraints involve time requirements and shift sequences. Therefore, the aim is to meet the hard constraints while satisfying the soft constraints as much as possible. Tables 8.3 and 8.4 describe typical hard and soft constraints, respectively.

The next section presents the fuzzy particle swarm optimization algorithm proposed in this chapter.

TABLE 8.3

Typical Hard Constraints in Physician Scheduling

No.	Description	References
H1	The overall schedule should meet the minimum man-power requirements in every shift of every day	Ernst et al. (2004a,b); Puente et al. (2009)
H2	The total working hours of a physician cannot exceed 48 hours every week	Lo and Lin (2011)
H3	The total working hours of a physician cannot be less than 16 hours every week	Lo and Lin (2011)
H4	Consecutive working days of a physician cannot exceed 6 days	Lo and Lin (2011)
H5	After a night shift, a physician should have at least a day off	Puente et al. (2009); Lo and Lin (2011)
H6	Consecutive off days of a physician cannot exceed 5 days	Lo and Lin (2011)

TABLE 8.4

Typical Soft Constraints in Physician Scheduling

No.	Description	References
S1	The number of forbidden shift sequences should be as low as possible	Puente et al. (2009)
S2	The number of uncongenial shift allocations for each physician should be as low as possible	Mutingi and Mbohwa (2013)
S3	The variation of the number of each shift type assigned to an individual should be as low as possible	Mutingi and Mbohwa (2013)
S4	The variation of individual workloads should be as minimal as possible	Lo and Lin (2011)
S5	The number of temporary staff assignments should be minimized as much as possible	Puente et al. (2009)

8.4 FUZZY PARTICLE SWARM OPTIMIZATION APPROACH

Fuzzy particle swarm optimization (FPSO) is a development from the basic particle swarm optimization (PSO) algorithm formerly proposed by Eberhart and Kennedy (1995). The FPSO is enhanced by fuzzifying one or more procedures of the basic PSO, introducing domain-specific heuristics that supervise population initialization and shift assignment, and introducing an improved coding scheme.

Basically, the PSO algorithm is a population-based stochastic optimization inspired by social behavior of bird flocking or fish schooling. Potential candidate solutions, called particles, fly through the solution space by learning from and following the current best particles to move toward the best solution. The dynamics of the algorithm is influenced by four basic parameters: the size of the particle swarm, the maximum number of iterations, the local learning factors or confidence coefficients (c_1 and c_2), and the inertia weight (w). At every iteration, the position of each particle in the swarm is determined by its particle velocity. Mathematically, the velocity of particle p in dimension d, denoted by v_{pd}, is determined by the expression

$$v_{pd}(t+1) = w \cdot v_{pd}(t) + c_1 \cdot \text{rand}() \cdot [pbest_{pd}(t) - x_{pd}(t)]$$

$$+ c_2 \cdot \text{rand}() \cdot [gbest(t) - x_{pd}(t)] \tag{8.1}$$

where
  x_{pd} is the position of particle i in dimension d
  rand() is a random number generator that generates random values in [0 1]
  $pbest_{pd}$ denotes the individual optimum for particle p
  $gbest$ denotes the global optima for the entire particle swarm
  Inertia weight w represents the effect of previous velocity on the new velocity vector

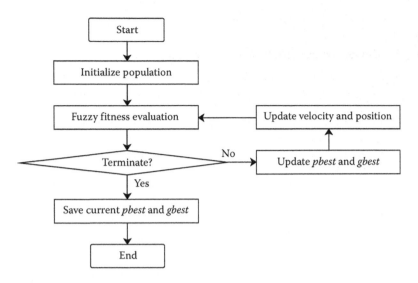

FIGURE 8.1   Proposed FPSO algorithm.

To further explain the FPSO, the algorithm is presented in terms of its constituent elements, including particle position representation scheme, initialization, fuzzy fitness evaluation, and velocity update, in this section. Figure 8.1 shows the flowchart of the FPSO procedure.

8.4.1   FPSO Coding Scheme

The FPSO coding scheme used in this research is derived from past nurse rostering problems, that is, nurse-day view and nurse-shift view (Cheang et al., 2003). Figure 8.2 shows an example of a candidate solution (a complete schedule) for nine physicians, spanning a period of 7 days. Shift types {D, E, N, W, H}, described in the previous section, are assigned to each physician on each day, where a blank represents a day off. A candidate solution is then encoded as a string of shifts, beginning from physician P1 up to P8. Figure 8.3 illustrates the resulting code for the problem.

Days / Physician	Mon	Tue	Wed	Thur	Fri	Sat	Sun
P1	D		D	D	D	H	
P2	D	D	E	E	N		
P3	E	D	E	D		D	D
P4	D		W	E	E		H
P5	N			D	E	E	E
P6	W	N	N	E		E	
P7		E	D	D	D		N
P8	E	D	N	N			H
P9	D	D	D	D	D	N	

FIGURE 8.2   The FPSO coding scheme.

	P1				P2							P8				
D		D	D	D	H	D	D	E	E	N		···	D	D	D	D	D	N

FIGURE 8.3 An example of a coding scheme for a candidate solution.

The proposed coding scheme ensures that each physician is assigned at most a single shift for each day.

8.4.2 INITIALIZATION

Given that there are a number of hard constraints, random generation of shifts may result in many infeasible candidate solutions. To prevent infeasibilities, corrective constructive techniques are included in the FPSO procedure. In this case, any infeasible shift assignment is rejected, should a hard constraint be violated. Soft constraints are formulated using fuzzy membership functions. These techniques help to improve the effectiveness and efficiency of the proposed FPSO algorithm.

In generating candidate solutions, real-valued or integer particle position values (coordinates) may be created. Basically, these particle position values are randomly created using the following expression:

$$x_i = X_{min} + (X_{max} - X_{min}) \times U(0,1) \qquad (8.2)$$

where X_{min} and X_{max} are the lower and upper limits of a predefined range of position values and $U(0,1)$ is a uniform random number in the range [0,1].

In this application, continuous position values in the range [0,1] are converted into shifts codes {M, A, N, D, H}. The generated position values are mapped to the nearest integer number using a suitable rounding function, round(), as follows:

$$x_i' = \text{round}(x_i) \qquad (8.3)$$

The FPSO algorithm generates an initial flock, where each bird is called a particle. Each particle flies at a certain velocity, to find a global best position, after a number of iterations. In each iteration, each particle adjusts its velocity according to its momentum, its best position (called *pbest*), and that of its neighbors (called *gbest*). This ultimately influences the particle's new position. Therefore, given a search space d and the total number of particles n, the position of the ith particle is $x_i = [x_{i1}, x_{i2},...,x_{id}]$, the best position of the ith particle is $pbest_i = [pbest_{i1}, pbest_{i2},...,pbest_{id}]$, and the velocity of the ith particle is $v_i = [v_{i1}, v_{i2},...,v_{id}]$.

In this chapter, an enhanced initialization algorithm is designed to satisfy all hard constraints of the NSP problem. Figure 8.4 presents the pseudocode for the proposed algorithm. The procedure begins by allocating leave days and holidays. This is followed by random assignment of the {D, E, N, W, H} shifts, while ensuring subsequent shift assignments do not belong to the forbidden set F comprising illegal shift sequences. In the actual implementation shift set, {D, E, N, W, H} are replaced by real numbers in [0,1].

Initialization algorithm
1.   Assign holidays, requested leave days;
2.   **Repeat**
3.   Randomly generate an initial shift k_1;
4.   **Repeat**
5.   Randomly generate shift k_n = rand(D, E, N, W, H);
6.   **If** sequence (k_{n-1}, k_n) ∉ Forbidden set F **Then**
7.   Add shift k_n to shift pattern P_i;
8. $n = n + 1$;
9.   **End If;**
10.   **Until** (Shift Pattern P_i is complete);
11.   **Until** (Required Shift Patterns, I, are generated);
12.   **Repeat**
13.   Compare number of assigned shifts and coverage requirements for each shift;
14.   **If** a shift k_n is understaffed **Then**
15.   Replace overstaffed shifts with shift k_n;
16.   **Until** (All shift requirements are met);

FIGURE 8.4   Enhanced FPSO initialization algorithm.

Following initialization, FPSO loops through fuzzy evaluation, velocity, and position update until the given termination criterion is satisfied.

8.4.3   FITNESS EVALUATION

The fitness of a solution is evaluated as the degree to which it satisfies the soft constraints converted to objective functions. Therefore, fitness is formulated as a function of the weighted sum of the satisfaction of each of the soft constraints. Furthermore, we represent each soft constraint as a normalized fuzzy membership function whose values fall in the range [0,1].

Linear membership functions, such as triangular and trapezoidal membership functions, have widely been recommended (Sakawa, 1993). Therefore, linear functions are used to define the fuzzy membership functions of the soft constraints. In this vein, let a and b denote the minimum and maximum of the feasible values of each objective function, and μ_h denote the membership function for objective h. Then, the membership functions corresponding to minimization and maximization can be defined based on the satisfaction degree, as illustrated in Figure 8.5.

For minimization, the linear membership function can be formulated according to the following expression:

$$
\mu_h = \begin{cases} 1 & z \leq b \\ \dfrac{b-z}{b-a} & a \leq z \leq b \\ 0 & z \geq b \end{cases} \tag{8.4}
$$

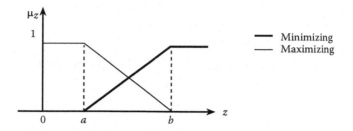

FIGURE 8.5 Interval-valued linear membership function.

Clearly, the function μ_h is monotonically decreasing in z. On the other hand, for the case of maximization, the membership function can be defined as follows:

$$\mu_h = \begin{cases} 1 & z \geq b \\ \dfrac{b-z}{b-a} & m_t \leq z \leq b \\ 0 & z \geq a \end{cases} \qquad (8.5)$$

Again, it can be seen that the function μ_h is a monotonically increasing function of z.

8.4.3.1 Membership Functions

To model physician preferences into our scheduling algorithm, fuzzy membership functions are used to measure satisfaction of soft constraints. For instance, a schedule [D D D D N] with a shift sequence variation of 1 is more desirable than [D N D N D] with a variation of 4. Therefore, shift sequence variation should be minimized as much as possible. In addition, the variation of individual physicians' workloads from the mean should be as low as possible. Other functions can be modeled in a similar manner, depending on the constraints pertinent to the problem domain. Table 8.5 lists the membership functions studied in this work.

8.4.3.2 The Overall Fitness Function

The resulting fitness is calculated as a weighted sum of the satisfaction of each of the soft constraints. Therefore, the final fitness is a function of the normalized membership functions, as shown by the expression

$$z = \sum_f w_f \mu_f \qquad (8.6)$$

where w_f is the weight of each function f, such that $\Sigma w_f = 1.0$. The weight parameter w_f enables the decision maker to model his or her choices to reflect the preferences and desires of the physicians and management. This gives FPSO an added advantage over related heuristics such as PSO and SA.

TABLE 8.5
Description of Membership Functions

No.	Membership	Fuzzy Parameter	Brief Description
1	μ_1	Shift variations in a shift sequence	Schedules like [D D D D N] with a shift sequence variation of 1 are more desirable than [D N D N D] with a variation of 4.
2	μ_2	Forbidden shift sequences	The number of forbidden shift sequences should be as low as possible.
3	μ_3	Congeniality	The number of uncongenial shift allocations for each physician should be as low as possible.
4	μ_4	Shift fairness	The variation of the number of each shift type assigned to an individual from the mean should be as low as possible.
5	μ_5	Workload fairness	The variation of each individual workload should be as minimal as possible.
6	μ_6	Temporary staff assignments	The number of temporary staff assignments should be minimized as much as possible.

8.4.4 VELOCITY AND POSITION UPDATE

Iteratively, the position and velocity at iteration $(t + 1)$ are updated according to the following:

$$v_i(t + 1) = w \cdot v_i(t) + c_1 \cdot \eta_1 \cdot (pbest_i(t) - x_i(t)) + c_2 \cdot \eta_2 \cdot (gbest(t) - x_i(t)) \quad (8.7)$$

$$x_i(t + 1) = x_i(t) + v_i(t + 1) \quad (8.8)$$

where
    c_1 and c_2 are constants
    η_1 and η_2 are uniformly distributed random variables in [0,1]
    w is an inertia weight showing the effect of previous velocity on the new velocity vector

8.4.5 FPSO TERMINATION CRITERIA

The termination of the FPSO procedure is controlled in two ways, that is, the algorithm terminates (1) when a preset maximum number of iterations are exceeded, and/or (2) when 100 iterations are executed without a significant improvement in the current best solution.

8.5 THE FPSO IMPLEMENTATION

Figure 8.6 shows the overall procedure of the FPSO algorithm, consisting of the procedures described in the previous section, that is, initialization, fuzzy fitness evaluation, and velocity and position update.

The FPSO algorithm

1. Input w, η_1, η_2, c_1, c_2, N;
2. Enhanced Initialization;
3. **For** $i = 1$ to N:
4.   Initialize particle position $x_i(0)$ and velocity $v_i(0)$;
5.   Initialize $pbest_i(0)$;
6. **End For;**
7. Initialize $gbest(0)$;
8. **For** $i = 1$ to N:
9.   Compute fuzzy fitness $f(x)$, $x = (x_1, x_2, ..., x_N)$;
10. **Repeat**
11.   **For** $i = 1$ to N:
12.   Compute fitness f_i;
13.   **If** (f_i > current *pbest*) then
14.   Set current value as new *pbest*;
15.   **If** (f_i > current *gbest*) then
16.   *gbest* = i;
17.   **End If;**
18.   **End If;**
19.   **End For;**
20.   **For** $i = 1$ to N:
21.   Find neighborhood best;
22.   Compute particle velocity $v_i(t + 1)$;
23.   Update particle velocity $x_i(t + 1)$;
24.   **End For;**
25. **Until** (termination criteria are fulfilled);

FIGURE 8.6 A pseudocode for the overall PSO-based algorithm.

The FPSO approach has a number of advantages. First, the structure of the algorithm is intuitively easy to follow and implement in a number of problem situations, with few adjustments. In addition, the algorithm is computationally efficient and effective, as it is able to obtain optimal or near-optimal solutions within a reasonable computation time. Moreover, fuzzy evaluation enables the algorithm to pass through inferior intermediate solutions, avoiding infeasibilities, which may eventually yield good solutions. Thus, instances of infeasible solutions are avoided during algorithm execution.

The next section presents illustrative computational experiments together with the experimental setups.

8.6 COMPUTATIONAL EXPERIMENTS

Sample problems were created from scheduling scenarios with 10, 14, and 20 physicians, to be scheduled over a period of 2 weeks. The usual constraints (hard and soft) presented in this study were assumed. Through preliminary tests and fine-tuning, the parameters of the FPSO algorithm were set up as in Table 8.6 Various parameter value combinations were tested for 20 runs for different problem sizes.

TABLE 8.6

Parameters of Soft Constraints

Parameter	Value
Population size	20
Number of iterations	300
$c_1{:}c_2$	1:1; 1:2; 2:1
w	0.6; 1.1

TABLE 8.7

Weightings of Soft Constraints

Constraint	S1	S2	S3	S4	S5
Weight	0.2	0.05	0.25	0.3	0.2

The weights of the soft constraints previously listed in Table 8.4 are normally obtained through consultations with management and the physicians. In this study the weights in Table 8.7 were used.

The next section presents the results and discussions from the previously described computational experiments.

8.7   RESULTS AND DISCUSSIONS

The primary experiment consists of 14 physicians to be scheduled over 14 days, as depicted in Figure 8.7. Shaded columns represent nonworking days. Hard constraints

Physician \ Day	1	2	3	4	5	6	7	8	9	10	11	12	13	14
P1	D		N		N			A	N		D		H	
P2		N		D				N		N		D		
P3	A	A	A	A	A			M	M	M	M	M	H	
P4	A	A	A	A	A			D		N		N		
P5	M	M	M	M	M	H		A	A	A	A	A		
P6		N		N				N		D		N		
P7	M	M	M	M	M			A	A	A	A	A	H	
P8	N		D		N			M	M	M	M	M		H
P9	A	A	A	A	A	H			N		N		H	
P10	M	M	M	M	M			A	A	A	A	A		
P11	A	A	A	A	A		H		A	A	A	A		
P12	M	M	M	M	M	H		M	D		N			
P13	N		N		D			M	M	M	M	M		
P14		D		N	N		H		M	M	M	M		

FIGURE 8.7   A good intermediate solution of the FPSO algorithm.

TABLE 8.8
Soft Constraint Satisfaction

Constraint	S1	S2	S3	S4	S5
Weight	0.2	0.05	0.25	0.3	0.2
Satisfaction	100%	93.8%	100%	100%	96%
p-Value (99%)	<0.05	<0.05	<0.05	<0.05	<0.05

TABLE 8.9
FPSO Performance with Varying Problem Sizes

Experiment	Problem Size	CPU Time (seconds)
1	10	62.48
2	14	63.44
3	20	68.24
4	25	70.12
5	30	71.04

are always satisfied, as expected. However, it is desirable that each soft constraint be satisfied as much as possible, possibly up to 100% satisfaction.

The fitness of the final solution is subject to the user's approval. It is important to show to a certain level of confidence, the degree of satisfaction of each soft constraint, considering the given weightings. It is assumed that each constraint is adequately satisfied to the weighting assigned to the constraint. Table 8.8 presents the results of constraint satisfaction, given a level of confidence of 99%, based on a one-tail hypothesis test. Therefore, all the targets for the constraints were fully fulfilled, showing the effectiveness of the algorithm.

Table 8.9 shows a comparative analysis of the performance of the FPSO algorithm over a varying number of physicians (problem size). FPSO performance is measured in terms of average CPU time, in seconds, for 20 runs of each experiment. It can be seen that as problem size is increased from 10 to 25, performance increases marginally from 62.48 to 71.04. Therefore, the effect of problem size on algorithm performance is not significant. This reveals that the algorithm is effective and efficient even for large-scale problems.

8.8 SUMMARY

The physician scheduling problem is a complex problem characterized by uncertain and unpredictable demand and complex work regulations. High-quality schedules are necessary to improve worker moral, thus avoiding absenteeism and attrition. In an environment where physician preferences are ill-defined, the use of fuzzy set theory concepts is a suitable option. Developing efficient and effective multicriteria decision methods is crucial. It is necessary to capture uncertain preferences, goals, and constraints of the problem. Multicriteria heuristics can address the physician scheduling problem. This research proposed an FPSO algorithm with a fuzzy goal-based fitness function for solving the physician scheduling problem. An enhanced solution generation heuristic is developed for better efficiency. Computational experimental results showed that the algorithm is capable of solving large-scale scheduling problems. Therefore, the approach provides a useful contribution to academicians and to practitioners in healthcare service organizations.

8.8.1 CONTRIBUTIONS TO KNOWLEDGE

The FPSO algorithm proposed in this study contributes to the healthcare operations management community. It provides an approach to solving physician scheduling problems characterized with imprecise preferences and management goals that are often expressed as soft constraints. By using fuzzified heuristics, the approach incorporates more realism into the solution process. As opposed to conventional linear programming methods, the algorithm is capable of handling large-scale problems, while able to provide optimal or near-optimal solutions within a reasonable computation time. Therefore, the metaheuristic is a good basis for decision support to decision makers in the healthcare sector. Moreover, the method is also a useful contribution to the practicing decision maker.

8.8.2 CONTRIBUTIONS TO PRACTICE

In practice, the proposed algorithm can offer the user an opportunity to use weights to model preferences and choices in an interactive manner, which is rather impossible with crisp methods. Decision makers prefer to use interactive approaches that provide a population or a list of good alternative solutions, rather than those that prescribe a single "optimal solution." Interactively, the user can use information from physicians and other members of management (coded as weights) to make judicious adjustments to candidate solutions. Therefore, the use of FPSO can be more acceptable to most practicing decision makers in the field. The FPSO method is an effective and efficient algorithm for physician scheduling problems.

REFERENCES

Aickelin, U. and Dowsland, K. 2004. An indirect genetic algorithm for a nurse scheduling problem. *Computers and Operations Research* 31: 761–778.
Bailey, J. 1985. Integrated days off and shift personnel scheduling. *Computers & Industrial Engineering* 9 (4): 395–404.
Bailey, R. N., Garner, K. M. and Hobbs, M. F. 1997. Using simulated annealing and genetic algorithms to solve staff scheduling problems. *Asia-Pacific Journal of Operational Research* 14: 27–43.
Bard, J. F. and Purnomo, H. W. 2005. Preference scheduling for nurses using column generation. *European Journal of Operational Research* 164 (2): 510–534.
Beaumont, N. 1997. Scheduling staff using mixed integer programming. *European Journal of Operational Research* 98: 473–484.

Bergh, J., Beliën, J., Bruecker, P., Demeulemeester, E. and Boeck, L. 2013. Personnel scheduling: A literature review. *European Journal of Operational Research* 226: 367–385.

Bester, M. J., Nieuwoudt, I. and Van Vuuren, J. H. 2007. Finding good nurse duty schedules: A case study. *Journal of Scheduling* 10 (6): 387–405.

Burke, E., Kendall, G. and Soubeiga, E. 2003. A tabu search hyperheuristic for timetabling and rostering. *Journal of Heuristics* 9: 451–470.

Burke, E. K., Causmaecker, P., Petrovic, S. and Berghe, G. V. 2006. Metaheuristics for handling time interval coverage constraints in nurse scheduling. *Applied Artificial Intelligence* 20 (9): 743–766.

Cheang, B., Li, H., Lim, A. and Rodrigues, B. 2003. Nurse rostering problems—A bibliographic survey. *European Journal of Operational Research* 151: 447–460.

Cheng, M., Ozaku, H.I. Kuwahara, N., Kogure, K. and Ota, J. 2007. Nursing care scheduling problem: Analysis of staffing levels. *Proceedings of the 2007 IEEE International Conference on Robotics and Biomimetics*, December 15–18, 2007, Sanya, China, 1: 1715–1719.

Daknou, A., Zgaya, H., Hammadi, S. and Hubert, H. 2010. A dynamic patient scheduling at the emergency department in hospitals. *IEEE Workshop on Health Care Management* (WHCM) 2010: 1–6.

Dowsland, K. A. 1998. Nurse scheduling with tabu search and strategic oscillation. *European Journal of Operations Research* 106: 393–407.

Eberhart, R. C. and Kennedy, J. 1995. A new optimizer using particles swarm theory. *Proceedings of the 1995 IEEE International Conference on Neural Networks* 4: 1942–1948.

Ernst, A. T., Jiang, H., Krishnamoorthy, M., Owens, B. and Sier, D. 2004a. An annotated bibliography of personnel scheduling and rostering. *Annals of Operations Research* 127 (1–4): 21–144.

Ernst, A. T., Jiang, H., Krishnamoorthy, M. and Sier, D. 2004b. Staff scheduling and rostering: A review of applications, methods and models. *European Journal of Operational Research* 153 (1): 3–27.

Ferrin, D. M., Miller, M. J., Wininger, S. and Neuendorf, M. S. 2004. Analyzing incentives and scheduling in a major metropolitan hospital operating room through simulation. In *Proceedings of the 2004 Winter Simulation Conference* 2: 1975–1980.

Gong, Y.-J., Zhang, J. and Fan, Z. 2010. A multiobjective comprehensive learning particle swarm optimization with a binary search-based representation scheme for bed allocation problem in general hospital. *IEEE International Conference on Systems Man and Cybernetics* (SMC), 2010: 1083–1088.

Gorunescu, F., McClean, S. I. and Millard, P. H. 2002. Using a queuing model to help plan bed allocation in a department of geriatric medicine. *Health Care Management Science* 5 (4): 307–312.

Hertz, A. and Lahrichi, N. 2009. A patient assignment algorithm for home care services. *Journal of the Operational Research Society* 60 (4): 481–495.

Kim, S. C., Horowitz, I., Young, K. K. and Buckley, T. A. 2000. Flexible bed allocation and performance in the intensive care unit. *Journal of Operations Management* 18 (4): 427–443.

Lin, W. S. and Yeh, J. Y. 2007. Using simulation technique and genetic algorithm to improve the quality care of a hospital emergency department. *Expert Systems with Applications* 32: 1073–1083.

Lo, C.-C. and Lin, T.-H. 2011. A particle swarm optimization approach for physician scheduling in a hospital emergency department. *7th International Conference on Natural Computation* 4: 1929–1933.

Mutingi, M. and Mbohwa, C. 2013. Home healthcare staff scheduling: A taxonomic state-of-the-art review. *IEEE International Conference on Industrial Engineering and Engineering Management*, Thailand, December 10–13, 2013, 1107–1111.

Mutingi, M. and Mbohwa, C. 2015. A multicriteria approach for nurse scheduling—Fuzzy simulated metamorphosis algorithm approach. *IEEE International Conference on Industrial Engineering and Operations Management*, March 3–5, 2015, Dubai (forthcoming).

Puente, J., Gómez, A., Fernández, I. and Priore, P. 2009. Medical doctor rostering problem in hospital emergency department by means of genetic algorithms. *Computers & Industrial Engineering* 56: 1232–1242.

Sakawa, M. 1993. *Fuzzy sets and interactive multi-objective optimization*, Plenum Press, New York.

Topaloglu, S. and Selim, S. 2007. Nurse scheduling using fuzzy multiple objective programming. In *New trends in applied artificial intelligence, lecture notes in computer science*, eds. H. G. Okuno and M. Ali, 4570: 54–63.

Topaloglu, S. and Selim, S. 2007. 2010. Nurse scheduling using fuzzy modeling approach. *Fuzzy Sets and Systems* 161: 1543–1563.

9 Fuzzy Grouping Genetic Algorithm for Homecare Staff Scheduling

9.1 INTRODUCTION

Homecare services provide special medical and paramedical assistance to patients in their homes (Bachouch et al., 2010). Homecare workers provide therapy, house cleaning, medical, and social services. Elderly people, patients with physical or mental challenges or acute illness, and those in need of posthospitalization treatment receive special care in their homes (Woodward, Tedford, and Hutchison, 2004; Akjiratikarl, Yenradee, and Drake, 2007; Mutingi and Mbohwa, 2013). The demand for homecare services is aggravated by an ageing population and the ever-increasing chronic pathologies. Governmental authorities and communities continue to call for improved healthcare services (Drake and Davies, 2006). On the other hand, modern innovative technologies enhance provision of homecare services (Mutingi, 2014).

Caregivers visit the patients at their homes at specific time windows preferred by patients. For instance, Figure 9.1 shows three patients, P1 to P3, to be visited by a caregiver at specific time windows over a single trip within shift time 0800 hours and 1400 hours. Healthcare workers travel for some distance to deliver care to their assigned patients at their homes within a specified time window, and then return to their workplace. All tasks assigned to the caregiver have to be completed within his or her capacity limit—for instance, a caregiver's capacity limit of 8 hours per day (Drake and Davies, 2006). As the number of patients increases, the problem becomes complex. With large problems, it may not be possible to meet all the time window preferences. However, it is essential to minimize, as much as possible, cost functions such as time window violations, traveling costs, and workload variation.

To provide satisfactory homecare services, decisions in healthcare operations must be optimized while considering patient satisfaction, healthcare worker satisfaction, and target management goals. Patient visits must be done at time windows preferred by patients, workload imbalances and unfair task assignments should be avoided as much as possible, and long, costly distance trips to patients should be minimized. This calls for models that can handle fuzzy multiple criteria. In this vein, this chapter seeks to present a fuzzy multicriteria approach to address the homecare staff scheduling problem. The research work follows through on these objectives:

1. To describe the homecare staff scheduling problem with time windows
2. To propose a fuzzy grouping genetic algorithm approach for solving the problem
3. To carry out illustrative computational experiments using a proposed approach

119

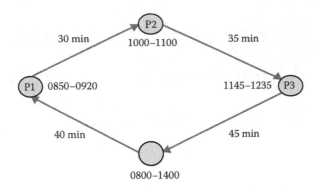

FIGURE 9.1   An example of a homecare staff schedule.

Advantages associated with efficient homecare scheduling are vast, for instance, (1) significant reduction of traveling costs of homecare staff; (2) improved worker utilization, leading to reduced labor costs; (3) improved healthcare services, satisfying patients and management goals; and (4) reduced scheduling time, freeing up the decision maker's time for strategic roles.

The next section presents a review of related literature. The homecare staff scheduling problem is then described. This is followed by a description of the proposed fuzzy grouping genetic methodology. Illustrative computational experiments are presented. A summary and further research prospects conclude the chapter.

9.2   RELATED WORK

Homecare staff scheduling has attracted significant attention in recent years. Researchers have applied various methods, including, mathematical programming (MP) models, heuristics, and metaheuristics. Bachouch et al. (2010) developed an MP-based optimization model for task assignment in a home health center. The mixed-integer program was solved using MS Excel and LINGO solver. Borsani et al. (2006) presented a mathematical model based on integer linear programming techniques. Heuristic approaches are more attractive, especially over large-scale problems. Bertels and Fahle (2006) presented a hybrid approach combining linear programming, constraint programming, and heuristics, to minimize transportation costs while maximizing satisfaction of patients and care providers. Eveborn, Flisberg, and Ronnqvist (2006) introduced a homecare scheduling problem for a variety of care providers and formulated the problem using a set partitioning approach. Akjiratikarl, Yenradee, and Drake (2007) developed a particle swarm optimization (PSO)-based algorithm for homecare staff scheduling problems in the United Kingdom.

Cheng and Rich (1998) modeled homecare staff scheduling as a vehicle routing problem with time windows (VRPTW) and multiple depots. Begur, Miller, and Weaver (1997) developed a decision support system using simplified scheduling heuristics. From an MIP model, the authors proposed a heuristic approach to minimize labor costs. However, though a number of approaches exist in the literature, designing more efficient and effective methods is imperative. Interactive methods that offer decision makers a population of good alternative solutions are preferable—for example, genetic algorithms (GAs) (Goldberg, 1989) and PSO (Kennedy and Eberhart, 1995). Given the increasing variability of the characteristics of problem instances, flexible and adaptable methods that can exploit and model specific problem structures and intuitive choices of the decision maker are also favorable. This chapter addresses these voids.

9.3   THE HOMECARE STAFF SCHEDULING PROBLEM

The homecare staff scheduling problem (HSSP) can be described as follows. Consider a team of m caregivers, who are to visit n patients, from a healthcare center. Each caregiver k ($k = 1,2,...,m$) is supposed to visit a group of patients. Each patient j ($j = 1, 2,...,n$) is supposed to be visited within a specific time window defined by earliest and latest start times, e_j and l_j, respectively. The aim is to minimize costs incurred by each caregiver in traveling from the point of origin to the patients and back to the origin. Additionally, penalty costs incurred when a caregiver arrives at the patient earlier than e_j or later than l_j must be minimized. Let a_j be the caregiver's arrival time at patient j, and c_e and c_l denote the unit penalty costs incurred when the caregiver arrives too early or too late, respectively. Consequently, expressions $\max[0,e_j - a_j]$ and $\max[0,a_j - l_j]$ have to be minimized, which is equivalent to minimizing time window violation, leading to patient satisfaction. Care workers prefer fair workload assignments. This problem is analogous to the vehicle routing problem with time windows (Li and Guo, 2001; Xiaoxiang, Weigang, and Jianwen, 2009). The next section describes the fuzzy grouping genetic algorithm for the scheduling problem.

9.4   THE FUZZY GROUPING GENETIC ALGORITHM APPROACH

The fuzzy grouping genetic algorithm (FGGA) is a development from the grouping genetic algorithm (GGA) originally developed by Falkenauer (1992). The GGA was developed to address problems whose candidate solutions can be coded into a group-structured chromosome. The FGGA encodes the group structure of the HSSP and incorporates fuzzified techniques into the grouping genetic operators. Unique enhanced genetic operators go a long way in strengthening the algorithm. A flowchart of the FGGA is shown Figure 9.2.

9.4.1   GROUP ENCODING SCHEME

The efficiency and effectiveness of the FGGA is strongly influenced by the structure of the genetic coding (Filho and Tiberti, 2006; Mutingi and Mbohwa, 2012). A unique coding scheme is developed in this study aimed at exploiting the group structure of the scheduling problem. In this vein, suppose that $C = [1, 2, 3,...,n]$ is a chromosome representing a set of n patients to be visited by m caregivers. The evaluation of C involves partitioning the patients along C into m groups such that the desired goals and preferences are satisfied.

To illustrate, assume a hypothetical situation of a homecare staff scheduling problem consisting of $m = 3$ workers ($k = 1,2,3$) and $n = 7$ patients ($j = 1,2,...,7$). Assume

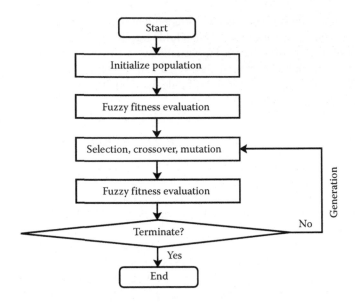

FIGURE 9.2 Flowchart of fuzzy grouping genetic algorithm.

that each of the 3 workers is to be assigned a set of patients to visit. One of the many possible solutions is to assign care workers 1, 2, and 3 to groups of patients {1,2}, {3,4,5}, and {6,7}, respectively. The group structure of the candidate solution can be represented as shown in Figure 9.3. The first part, part (a) provides the network structure of the problem while part (b) illustrates the group structure of the chromosome consisting of two codes. Code 1 represents the assignment of care workers s_1, s_2, and s_3, to groups of patients {1,2}, {3,4,5}, and {6,7}, respectively. Code 1 is the actual group structure upon which the genetic operators act. Code 2 denotes the last position of each patient group; that is, it records the position of the delimiter or frontier "|" that separates patient groups. For the example at hand, the cost in terms

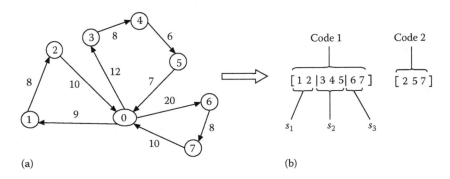

FIGURE 9.3 FGGA chromosome coding. (a) Network structure of the problem. (b) Group chromosome structure.

of distance traveled for each nurse is $(9 + 8 + 10)$, $(12 + 8 + 6 + 7)$, and $(20 + 8 + 10)$, respectively.

9.4.2 INITIALIZATION

Once the coding has been defined, the FGGA approach must initialize the population. An initial population of the desired size, *popsize*, is randomly created by random assignments of patients to caregivers. First, the care duties are arranged in ascending order of their start times. In case of a tie, the duties are ranked according to their activity duration. For each caregiver, a duty is assigned at a probability, starting from the earliest. From the unassigned set of duties, duties are assigned beginning from the earliest. This procedure increases the likelihood of the initialization process generating initial feasible solutions.

9.4.3 FUZZY FITNESS EVALUATION

The nature of the evaluation function is very crucial to the success of the algorithm. Therefore, the function should measure the relevant quality of the candidate solution, capture the imprecise conflicting goals and constraints, and be easy to compute. It is important to express the desirability η_i of a candidate solution relative to each care worker i. Due to the presence of multiple objective functions, the evaluation function η_i at each iteration should be expressed as a function of its constituent normalized functions, denoted by μ_h $(h = 1,...,q)$, where q is the number of objective functions. Therefore, we use a fuzzy multifactor evaluation method to determine the desirability, η_i, relative to worker i as follows:

$$\eta_i = \sum_h w_h \mu_h(z) \tag{9.1}$$

where z represents a specific objective function evaluation of a candidate solution at iteration t, and w_h is the weight of the function η_h, such that $\sum w_h = 1.0$.

The weight parameter w_h presents invaluable practical advantages. This parameter offers the decision maker an opportunity to incorporate choices or preferences of management and the nursing staff. Moreover, the use of the max–min operator is avoided so as to prevent possible loss of vital information. In this manner, the approach provides the FGGA algorithm an advantage over other metaheuristic approaches.

The fitness or quality of a solution is a function of how much it satisfies the preferences of nurses and patients, as well as management goals. Due to ambiguity and imprecision of these preferences and goals, fitness is modeled as normalized interval-valued fuzzy membership functions, as shown in Figure 9.4. For instance, management desires to minimize, as much as possible, the cost due to distance traveled. Thus, management may assert that the most desirable cost should be in the range $[0, a]$, beyond which the desirability linearly decreases down to 0 when cost reaches a limit b. The desirability is expressed as a membership function in Figure 9.4.

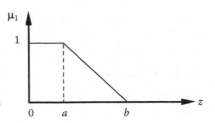

FIGURE 9.4 A trapezoidal fuzzy membership function.

Therefore, the corresponding membership function is represented by the following expression:

$$\mu_A(z) = \begin{cases} 1 & \text{If } z \leq a \\ (b-z)/(b-a) & \text{If } a \leq z \leq b \\ 0 & \text{If otherwise} \end{cases} \qquad (9.2)$$

In practice, the values of a and b are derived from expert knowledge and intuition. For instance, an average workload w is set as a standard or target workload such that the variation of individual workloads from w should at best be within the range $[0, a]$. In this case, a is the agreeable deviation of working time (hours) from the target w and considered unlikely to cause any dissatisfaction among the workers. To find b, one has to find the minimum of the maximum acceptable deviation of working time from the target. For example, if the choices of individual care workers ($i = 1,...,m$) on a and b are $a_1,...,a_m$ and $b_1,...,b_m$, respectively, then the values of a and b can be estimated:

$$a = \min(a_1,...,a_m) = a_1 \wedge a_2 \wedge...\wedge a_m \qquad (9.3)$$

and

$$b = \min(b_1,...,b_m) = b_1 \wedge b_2 \wedge...\wedge b_m \qquad (9.4)$$

where "$\wedge$" is the fuzzy min operator.

A similar approach as the preceding can be used to determine the values of a and b for the case of time window preferences and traveling distance in expressions. For time window preferences, individual choices of patients on a and b are considered. On the other hand, choices of a pool of decision makers can be used to estimate a and b for the traveling distance costs.

9.4.3.1 Membership Function 1: Earliness

To maximize satisfaction of the patients, their time windows must be satisfied. Earliness of nurse i is defined as $z_{1i} = [e_k - a_k]^+$, where, a_k is the actual visit time for patient k and e_k is the earliest acceptable service start time. It follows that earliness z_{1i} of nurse i must be minimized to an acceptable level:

$$\mu_4 = \mu_A(z_{1i}) \qquad (9.5)$$

where z_{1i} is the earliness of nurse i, and a and b are the fuzzy parameters for the function z_{1i}.

9.4.3.2   Membership Function 2: Lateness

Lateness of nurse i can be defined by $z_{2i} = [a_k - l_k]$, where, a_k is the actual visitation time for patient k and l_k is the latest acceptable service start time. Therefore, z_{2i} must be minimized to an acceptable level. Assuming the interval-valued membership function,

$$\mu_3 = \mu_A(z_{2i}) \qquad (9.6)$$

where z_{2i} is the lateness of nurse i, and a and b are the fuzzy parameters for the function z_{2i}.

9.4.3.3   Membership Function 3: Overload

For a fair workload assignment, overload (above the mean workload) should be minimized. Therefore, given workload h_i for each nurse i and mean workload w, the overload $z_{3i} = [h_i - w]^+$ should be minimized satisfactorily. Assuming an interval-valued membership function, we obtain

$$\mu_3 = \mu_A(z_{3i}) \qquad (9.7)$$

where z_{3i} is the overload for nurse i from the mean w; a and b are the fuzzy parameters for z_{3i}.

The same argument holds for any other cost functions, such as underload, distance cost, and overtime cost. Table 9.1 presents a descriptive summary of the membership functions in this study.

TABLE 9.1

Description of Membership Functions

No.	Membership	Fuzzy Parameter	Brief Description
1	μ_1	Earliness	Earliness of each nurse in visiting each assigned patient should be as low as possible to ensure acceptable service start time.
2	μ_2	Lateness	Lateness of each nurse in visiting each assigned patient should be as low as possible to ensure acceptable service start time.
3	μ_3	Overload	For a fair workload assignment, overload (above the mean workload) should be minimized as much as possible.
4	μ_4	Underload	For a fair workload assignment, underload (below the mean) should be minimized as much as possible.
5	μ_5	Distance cost	The cost is incurred when nurse i visits from point of origin 0, to all assigned patients, and back to the origin.
6	μ_6	Overtime cost	Overtime costs must be minimized as much as possible—that is, overtime must be within acceptable limits.

To evaluate the fitness of each candidate solution, our FGGA maps the objective function to a fitness function F_k, according to the expression

$$F_k(t) = \max [0, z^m(t) - z_k(t)] \tag{9.8}$$

where $z_k(t)$ is the objective function of candidate solution k at iteration t and z^m is the maximum objective function in the current population.

Multifactor evaluation is carried out for each factor. For each nurse i we obtain the membership function η_i, which is expressed as a weighted sum of μ_1, $\mu_1,...,\mu_5$ representing the desirability of a schedule in terms of workload balance, time window violation, and cost. Consequently, the membership function η_i for each worker i is given by

$$\eta_i = \sum_{h=1}^{5} w_h \mu_h \tag{9.9}$$

To obtain the overall fitness F_t for each candidate solution, the fuzzy min operator "∧" is used:

$$F_t = \eta_1 \wedge \eta_2 \wedge ... \wedge \eta_m \tag{9.10}$$

9.4.4 CROSSOVER

Crossover is one of the most important operators in grouping genetic algorithms. Basically, it is a mechanism thorough which selected chromosomes mate to produce new offspring, called selection pool. This enables the FGGA to explore unvisited regions in the solution space, which essentially provides the algorithm with explorative search abilities. Groups of genes in the selected chromosomes are exchanged at a probability p_c. First, a crossover point c is randomly generated between 1 and g, where g is the number of groups. Second, the group identified by c is swapped. Third, the offspring are repaired, if necessary. This process is repeated till the desired pool size, *poolsize*, is achieved. Figure 9.5 is an illustration of the crossover mechanism using parent chromosomes P1 and P2. Upon crossover, offspring O1 and O2 are

Parents:

P1: [5 2 | 4 3 1 | 6]

P2: [6 5 | 3 1 | 4 2]

Offspring:

O1: [5 2 | 3 1 | 6]

O2: [6 5 | 4 3 1 | 4 2]

Repaired:

O1′: [5 2 | 3 1 | 4 6]

O2′: [6 5 | 4 3 1 | 2]

FIGURE 9.5 Crossover and repair mechanisms.

produced, which are necessarily repaired since some genes are repeated and some are missing. After repair, offspring O1' and O2' are obtained.

After crossover, some genes may appear in more than one group, while others may be missing. Such offspring are repaired by (1) eliminating duplicated genes to the left and right of the crossover point, and (2) inserting missing genes into the groups with the least loading. Here, group coding takes advantage of the group structure to generate new offspring. The mutation operation follows the crossover operator.

9.4.5 MUTATION

The role of the mutation operator is to insert new characteristics into a population so as to enhance the search space of the FGGA. Mutation is applied to every new chromosome using two mutation operators: swap mutation and shift mutation. The swap mutation swaps genes between two groups in an individual chromosome, while the shift mutation works by shifting a randomly chosen frontier between two adjacent groups by one step, either to the right or to the left. In retrospect, the mutation operator essentially provides the FGGA with local search capability, a phenomenon called intensification. However, the shift mutation is a more localized search operation than the swap mutation. Figure 9.6(a) and (b) illustrates the swap and shift mutation mechanisms, respectively.

9.4.6 INVERSION AND DIVERSIFICATION

As iterations proceed, the population converges to a particular solution. However, population diversity has to be controlled in order to avoid premature convergence before an optimal solution is obtained, a process called genetic drift. Inversion is a genetic mechanism whose role is to propose the same solution to the FGGA, but differently, for the purpose of improving the diversity of the population at each generation or iteration.

Because crossover mechanisms work through crossing sites, the way in which a solution is presented influences the crossover operator's results; the first group appearing in the group element of a chromosome is less likely to be selected than the other groups. Therefore, it is important to include this operator in the FGGA procedure. For instance, groups of genes in a chromosome [1 2 | 4 | 3 5 6] are rearranged to [6 5 3 | 4 | 2 1], but at a very low probability.

Before mutation	:	[5 2	4 3 1	6]	[5 2	4 3 1	6]
After mutation	:	[5 2	6 3 1	4]	[5 2	4 3	1 6]
		(a)	(b)

FIGURE 9.6   (a) Swap and (b) shift mutation.

In order to check and control population diversity, first define an entropic measure, h_i, for each patient i according to the following expression:

$$h_i = \sum_{j=1}^{n} \frac{(x_{ij}/p) \cdot \ln(x_{ij}/p)}{\ln(n)} \tag{9.11}$$

where x_{ij} is the number of chromosomes in which patient i is assigned position j in the current population; n is the number of patients.

Therefore, diversity h can be defined according to the following expression:

$$h = \sum_{i=1}^{n} h_i/n \tag{9.12}$$

Inversion is applied whenever diversity falls below a threshold value, h_d. The best-performing candidates should always be preserved by comparing the diversified and undiversified populations and keeping a prespecified number of the best candidates in the population.

9.5 GGA OVERALL ALGORITHM

The overall structure of FGGA incorporates the operators presented in previous sections. Figure 9.7 presents an outline of the pseudocode for the overall structure of the GGA approach, consisting of all its constituent genetic operators.

The FGGA begins by selecting suitable input genetic parameters. In this study, the selected input genetic parameters were as follows: crossover probability (0.4), mutation probability (0.1), and inversion probability (0.02). Subsequently, an initial population, P(0), was then generated randomly by random assignments of patients to caregivers. The algorithm then proceeded into an iterative loop involving selection, group crossover, mutation, replacement strategy, inversion and diversification, population advancement, and termination condition test. The iterative loop allows successive iterations up until the number of iterations reaches a prespecified maximum, T, or when there is no significant improvement in the best solution in the population.

The proposed FGGA approach has a number of desirable advantages in its application. First, its procedure is easy to follow and therefore can be implemented easily in a number of problem situations. It is robust and versatile in that it is applicable to similar problems with little or no fine-tuning. Moreover, the algorithm is computationally efficient, obtaining good solutions within reasonable computation times. Most importantly, it incorporates fuzzy evaluation into the algorithm, allowing the global optimization process to pass through inferior solutions, which will eventually yield improved solutions. Fuzzy evaluation ensures that instances of infeasible solutions are avoided during the execution of the algorithm.

The FGGA overall algorithm	
1:	Input: FGGA parameters; $t = 0$;
2:	Initialize population, $P(0)$; Input FGGA parameters;
3:	**Repeat**
4:	Selection:
5:	Evaluate $P(t)$;
6:	Create temporal population, *temppop(t)*;
7:	Group crossover:
8:	Select 2 chromosomes from *temppop(t)*;
9:	Apply crossover operator;
10:	Repair if necessary;
11:	Mutation:
12:	Mutate $P(t)$;
13:	Add offspring to *newpop(t)*;
14:	Replacement:
15:	Compare successively, *spool(t)* and *oldpop(t)* strings;
16:	Take the ones that fare better;
17:	Select the rest of the strings with probability 0.55;
18:	Diversification:
19:	Calculate population diversity H;
20:	**While** $(h < h_d)$ THEN
21:	diversify $P(t)$;
22:	calculate h;
23:	**End While;**
24:	Evaluate $P(t)$;
25:	New population:
26:	oldpop(t) = newpop(t);
27:	Advance population, $t = t + 1$;
28:	**Until** (Termination criteria are satisfied);

FIGURE 9.7 Pseudocode for the overall GGA approach.

9.6 COMPUTATIONAL ILLUSTRATIONS

The proposed FGGA algorithm was coded and implemented in Java on a 3.06 GHz speed processor with 4GB RAM. Computational experiments and their results are presented. For the purpose of illustration, typical examples from the literature are adopted.

To demonstrate the performance of the algorithm, two categories of computational experiments are presented: (1) typical problem cases adapted from the literature, namely, problem case 1 and problem case 2; and (2) a set of 20 randomly generated problems, deriving from the typical problem cases in the first category. The second category demonstrates the effectiveness and efficiency of the algorithm when applied to large-scale problems.

The parameters of the FGGA were set as follows: population size of $p = 20$, crossover probability of $p_c = 0.35$, mutation probability of $p_m = 0.02$, and inversion probability $p_{in} = 0.04$. The stopping criteria were determined by the maximum number of iterations, $T_m = 300$, or the number of iterations without solution improvement,

$T_I < 30$. In addition, the weights were set to $w_h = 1$, for all $h = 1,2,...,5$. Each computational experiment was run 50 times.

9.6.1 CASE PROBLEM 1

The first problem case is an adaptation of a VRPTW instance presented in Liu, Weigang, and Jianwen (2009) and Li and Guo (2001). The VRPTW problem is mapped to the HSSP, where each healthcare staff represents a vehicle, and the patients represent the demand nodes. In this connection, assume that there are eight patients to be visited during their respective execution time windows, $[e_j, l_j]$, as shown in Table 9.2. Travel distances from the origin "0" to each patient (node) and between all other adjacent patients are provided in Table 9.3.

TABLE 9.2
Service Times and Time Windows[a]

Patient	Time Window [e_j, l_j]	Service Time (s_j)	Caregivers (b_j)
1	[60, 240]	60	1
2	[240, 360]	120	1
3	[60, 120]	60	1
4	[240, 420]	180	1
5	[180, 330]	60	1
6	[120, 300]	150	1
7	[300, 480]	180	1
8	[90, 240]	48	1

Source: Liu et al., 2009. *IEEE 2009 Fifth International Conference on Natural Computation*, pp. 502–506.

[a] Minutes.

TABLE 9.3
Distance between Patients (Demand Points)

	0	1	2	3	4	5	6	7	8
0	0	40	60	75	90	200	100	160	80
1	40	0	65	40	100	50	75	110	100
2	60	65	0	75	100	100	75	75	75
3	75	40	75	0	100	50	90	90	150
4	90	100	100	100	0	100	75	75	100
5	200	50	100	50	100	0	70	90	75
6	100	75	75	90	75	70	0	70	100
7	160	110	75	90	75	90	70	0	100
8	80	100	75	150	100	75	100	100	0

Source: Liu et al., 2009. *IEEE 2009 Fifth International Conference on Natural Computation*, pp. 502–506.

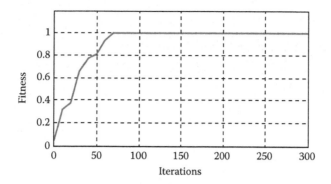

FIGURE 9.8 Illustrative computational runs based on problem case 1.

Furthermore, the original problem assumes that the unit travel cost is 1, the average traveling speed is 50, the penalty cost for violating any time window preference is $k_e = k_l = 50$, and the objective is to minimize the total costs due to traveling and violation of time window preferences. However, the aim of this study is to satisfy, as much as possible, the time window preferences, the workload balance, and the management goal on traveling costs.

Figure 9.8 provides an illustrative plot of the intermediate solutions arrived at in the iterative process of the FGGA algorithm. The plot is presented in steps of 10 iterations. The objective is to increase the fitness of the best solution as much as possible. It can be seen that the fitness value increased from 0.04 at the initial state to 1.00 at the 70th iteration, and it settled at that value. This means that, though the search process intended to run up to 300 iterations, the optimum solution was obtained at the 70th iteration, according to the plot.

The computational solution for the problem is presented in Table 9.4. The solution presented corresponds to the best-known result, with a total travel cost of $910 (Liu, Weigang, and Jianwen, 2009). The proposed FGGA was able to obtain the best-known solution, within an average of 3 seconds, which demonstrates the effectiveness of the method.

Table 9.5 shows a comparative analysis based on problem case 1. Comparison was done against the results previously obtained using competitive algorithms, such

---

TABLE 9.4

Computational Results for Problem Case 1

Staff	Route/Path
1	$0 \rightarrow 8 \rightarrow 5 \rightarrow 7 \rightarrow 0$
2	$0 \rightarrow 3 \rightarrow 1 \rightarrow 2 \rightarrow 0$
3	$0 \rightarrow 6 \rightarrow 4 \rightarrow 0$

Note: Fitness = 1.00; cost = $910.

TABLE 9.5

Comparative Analysis Based on Problem Case 1

Approach	Average Cost ($)	Success Rate (%)	Average CPU Time (s)	References
Basic GA	993.6	24	11.0	Li and Guo (2001)
PSO	940.5	46	6.00	Li, Zou, and Sun (2004)
Parallel PSO	923.8	72	4.00	Wu, Ye, and Ma (2007)
Hybrid PSO	914.0	97	4.00	Liu, Weigang, and Jianwen (2009)
FGGA	910.0	100	3.44	

as the basic GA, particle swarm optimization (PSO), parallel PSO, and hybrid PSO (Liu, Weigang, and Jianwen, 2009). The average computation time for the FGGA was 3.40 seconds, which is significantly less than that of all other algorithms. Moreover, the search success rate for the FGGA was 100%, which is far greater than other algorithms: 46% for PSO, 24% for the basic GA, 72% for parallel PSO, and 97% for hybrid PSO.

Overall, the comparative evaluations based on the average cost, search success rate time, and average computation times show that the proposed algorithm is effective and efficient in solving the HSSP. The next section presents the second problem case.

9.6.2 CASE PROBLEM 2

Problem case 2 was adopted from Trabelsi, Larbi, and Alouane (2012). Thus, the problem consists of $m = 3$ care workers and $n = 7$ patients. Table 9.6 provides data on the time window, $[e_j, l_j]$; the service time, s_j; and the required number of caregivers, b_j; for each patient j ($j = 1,...n$). Furthermore, Table 9.7 lists the travel times between patients, where "patient 0" represents the healthcare center, which is the origin for

TABLE 9.6

Patient Data

Patient	Time Window [e_j, l_j]	Service Time (s_j)	Caregivers (b_j)
1	[60, 90]	30	1
2	[300, 425]	55	1
3	[240, 280]	20	1
4	[180, 270]	30	2
5	[35, 95]	25	2
6	[30, 205]	35	1
7	[60, 220]	20	1

Source: Trabelsi, S. et al., 2012. In *BPM 2011 Workshops, Part II, LNBIP 100,* eds. F. Daniel et al., 143–151. Springer-Verlag, Berlin.

TABLE 9.7
Travel Times

	0	1	2	3	4	5	6	7
0	0	30	15	35	22	35	18	45
1	30	0	20	20	13	19	24	22
2	15	20	0	10	45	25	36	19
3	35	20	10	0	12	45	22	34
4	22	13	45	12	0	20	15	18
5	35	19	25	45	20	0	45	25
6	18	24	36	22	15	45	0	20
7	45	22	19	34	18	25	20	0

Source: Trabelsi, S. et al., 2012. In *BPM 2011 Workshops, Part II, LNBIP 100*, ed. F. Daniel et al., 143–151. Springer–Verlag, Berlin.

TABLE 9.8
Availability of Skills

Worker	1	2	3	4	5	6	7
1	1	1	0	1	1	0	1
2	1	0	1	1	1	1	1
3	1	1	1	1	1	1	0

Source: Trabelsi, S. et al., 2012. In *BPM 2011 Workshops, Part II, LNBIP 100*, ed. F. Daniel et al., 143–151. Springer–Verlag, Berlin.

each caregiver. Table 9.8 presents the availability of skills for the caregivers, where "1" denotes availability and "0" denotes unavailability of a specific skill for a particular caregiver. Computational results were compared with competing algorithms based on search success rate defined by the frequency of obtaining an optimal solution.

Table 9.9 shows the computational results for case problem 2. The FGGA algorithm obtained the best-known optimal solution in a few seconds. As in Trabelsi, Larbi, and Alouane (2012), both patients 4 and 5 received two arrivals of caregivers according to their stipulated requests.

Table 9.10 shows a comparative analysis of the performance of FGGA and GA, PSO, and linear programming (LP). The FGGA obtained an average fitness of 1.00, similar to the optimal solution obtained using LP (Trabelsi, Larbi, and Alouane, 2012). On the other hand, the fitness values for other algorithms are comparatively lower: 0.9 for GA and 0.93 for PSO. The average computation time for the FGGA was 2.86 seconds, which is less than that of other algorithms. Moreover, the success rate for the FGGA was 100%, as with LP (Trabelsi, Larbi, and Alouane, 2012), but comparatively higher than other algorithms, 92% for the GA and 93% for PSO. Therefore, the proposed method is efficient and effective.

TABLE 9.9

Computational Results for Case Problem 2

Worker 1		Worker 2		Worker 3	
Patient	Arrival Time	Patient	Arrival Time	Patient	Arrival Time
0	0800	0	0800	0	1038
5	0835	5	0835	4	1100
1	0930	7	1000	3	1230
4	100	6	1100	2	1320
0	1152	0	1152	0	1430

TABLE 9.10

Comparative Analysis of FGGA and Other Algorithms

Approach	Average Fitness	Search Success Rate (%)	Average CPU Time (s)
GA	0.92	92.0	4.52
PSO	0.93	94.00	3.62
LP[a]	1.00	100.0	3.00
FGGA	1.00	100.00	2.86

[a] Trabelsi, S. et al., 2012. In *BPM 2011 Workshops, Part II, LNBIP 100*, eds. F. Daniel et al., 143–151. Springer-Verlag, Berlin.

9.6.3 FURTHER EXPERIMENTS

Further experiments were randomly generated, with more than 20 patients and more than five care workers. Table 9.11 presents a summary of the results of the first eight problems, where m represents the number of care givers, n = number of patients, and $b_j = 1$ if only one care worker is required, and $b_j > 1$ if more than one care worker is required. Experimental results showed that the FGGA approach is capable of producing near-optimal solutions for large-scale problems within reasonable computation times. The computation times (CPU times in seconds) vary between 15 and 85 seconds for number of care workers $m = 4$, and between 22 and 119 seconds for $m = 5$. In general, computation times tend to increase with an increasing number of patients.

Figure 9.9 shows how the scheduling problem scales with the number of patients. It is interesting to see that the solution algorithm is largely independent of b_j. However, the algorithm is influenced by m, the number of caregivers. The less the available number of caregivers is, the more the computational effort before the algorithm settles for a near-optimal solution. In addition, the computational performance of the algorithm is influenced by n, the number of patients to be visited. The computational effort generally increases with the number of patients. However, it is interesting to note that the increase in computational effort is fairly gradual. This indicates the capability of the algorithm to solve large-scale problems efficiently.

TABLE 9.11
Further Computational Tests

m	n	CPU Time (s); $b_j = 1$	CPU Time (s); $b_j > 1$
4	7	15	15
	10	42	38
	15	62	66
	20	85	79
5	7	22	23
	10	81	75
	15	99	97
	20	114	119

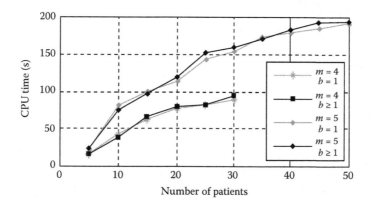

FIGURE 9.9   Comparative computational times (s).

Further increase in the problem size did not have a significant impact on the computation times, which gives the FGGA an advantage over large-scale problems. This suggests that the FGGA algorithm is capable of handling large-scale problem instances, unlike conventional linear programming approaches (Trabelsi et al., 2011). Moreover, the FGGA algorithm can efficiently address large-scale problems from a multiobjective perspective, providing near-optimal solutions.

9.7   SUMMARY

Homecare staff scheduling is a complex problem concerned with the construction of high-quality work schedules that satisfy (1) patient expectations in regards to their preferred time windows for home-based care, (2) staff preferences on workload assignments, and (3) management goals and choices. Constructing high-quality homecare schedules is imperative. This chapter proposed an FGGA optimization algorithm that utilizes the concepts of fuzzy set theory and a grouping genetic algorithm for solving typical homecare scheduling problems.

The fuzzy multicriteria evaluation method improved the performance of the algorithm in solving the multicriteria staff scheduling problem. In addition, the unique grouping operators can exploit and take advantage of the group structure of the problem. Moreover, the proposed algorithm benefits from its weight parameters that enable the decision maker to cautiously explore the most applicable trade-off between various conflicting decision criteria. Two categories of case problems were used to demonstrate the usefulness of the algorithm. The first category comprised case problems from the literature, while the second was randomly generated, deriving from the characteristics of the problems in the first category.

Computational results obtained showed that, compared to other related approaches, the FGGA algorithm is more efficient and effective in addressing the HSSP. The FGGA was able to optimally solve case problems in the literature with a success rate of 100%, an average fitness of 1.00, within computation times of a few seconds. Computational tests with large-scale problems showed that FGGA can produce near-optimal solutions within a range of 15 to 200 seconds. Therefore, the proposed FGGA is an efficient and effective metaheuristic approach to modeling homecare staff scheduling from fuzzy multiple criteria. In retrospect, the fuzzy grouping metaheuristic approach is a practical and flexible approach to HSSP.

It is interesting to extend the multicriteria fuzzy evaluation and grouping techniques employed to other complex problem domains.

REFERENCES

Akjiratikarl, C., Yenradee, P. and Drake, P. R. 2007. PSO-based algorithm for home care worker scheduling in the UK. *Computers & Industrial Engineering* 53: 559–583.

Bachouch, R. B., Liesp, A. G. Insa, L. and Hajri-Gabouj, S. 2010. An optimization model for task assignment in home healthcare. *IEEE Workshop on Health Care Management (WHCM)*, 1–6.

Begur, S. V., Miller, D. M. and Weaver, J. R. 1997. An integrated spatial decision support system for scheduling and routing home health care nurses. Technical report, Institute of Operations Research and Management Science.

Bertels, S. and Fahle, T. 2006. A hybrid setup for a hybrid scenario: Combining heuristics for the home health care problem. *Computers & Operations Research* 33 (10): 2866–2890.

Borsani, V., Matta, A., Beschi, G. and Sommaruga, F. 2006. A home care scheduling model for human resources. *International Conference on Service Systems and Service Management*, 449–454.

Cheng, E. and Rich, J. L. 1998. A home care routing and scheduling problem. Technical report, TR98-04, Department of Computational and Applied Mathematics, Rice University.

Drake, P. and Davies, B. M. 2006. Home care outsourcing strategy. *Journal of Health Organization and Management* 20 (3): 175–193.

Eveborn, P., Flisberg, P. and Ronnqvist, M. 2006. LAPS CARE—An operational system for staff planning of home care. *European Journal of Operational Research* 171: 962–976.

Falkenauer, E. 1992. The grouping genetic algorithms—Widening the scope of the GAs. *Belgian Journal of Operations Research, Statistics and Computer Science* 33: 79–102.

Filho, E. V. G. and Tiberti, A. J. 2006. A group genetic algorithm for the machine cell formation problem. *International Journal of Production Economics* 102: 1–21.

Goldberg, D. E. 1989. *Genetic algorithms in search, optimization and machine learning.* Addison-Wesley, Inc.: Boston.

Kennedy, J. and Eberhart, R. C. 1995. Particle swarm optimization. In *IEEE International Conference on Neural Networks 4*, Seoul, Korea, 1942–1948.

Li, J. and Guo, Y.-H. 2001. *Dispatch optimizes theory and methods for logistic distribution vehicle*. Chinese Commodity Publishing House: Beijing.

Li, N., Zou, T. and Sun, D. 2004. Particle swarm optimization for vehicle routing problem with time windows. *Systems Engineering-Theory & Practices*, 4: 130–135.

Liu, X., Weigang, J. and Jianwen, X. 2009. Vehicle routing problem with time windows: A hybrid particle swarm optimization approach. *IEEE 2009 Fifth International Conference on Natural Computation*, pp. 502–506.

Mutingi, M. 2014. Enhancing decision support in healthcare systems through mHealth. In *Mobile health (mHealth): Multidisciplinary verticals*, ed. S. Adibi, 508–522. CRC Press/Taylor & Francis: Boca Raton, FL.

Mutingi, M. and Mbohwa, C. 2012. Enhanced group genetic algorithm for the heterogeneous fixed fleet vehicle routing problem. *IEEM Conference on Industrial Engineering and Engineering Management*, December 10–13, 2012, Hong Kong, 207–211.

Mutingi, M. and Mbohwa, C. 2013. Home healthcare worker scheduling: A group genetic algorithm approach. *Proceedings of World Congress on Engineering 2013*, July 3–5, 2013, London, 721–725.

Trabelsi, S., Larbi, R. and Alouane, A. H. 2012. Linear integer programming for the home health care problem. In *BPM 2011 workshops, part II, LNBIP 100*, eds. F. Daniel et al., 143–151. Springer-Verlag, Berlin.

Woodward, C. J., Tedford, A. S. and Hutchison, B. 2004. What is important to continuity in home care? Perspectives of key stakeholders. *Social Science & Medicine* 58 (1): 177–192.

Wu, Y., Ye, C. and Ma, H. 2007. Parallel particle swarm optimization algorithm for vehicle routing problems with time windows. *Computer Engineering and Applications*, 43 (14): 223–226.

Xiaoxiang, L., Weigang, J. and Jianwen, X. 2009. Vehicle routing problem with time windows: A hybrid particle swarm optimization approach. *IEEE 2009 Fifth International Conference on Natural Computation*, 502–506.

10 Fuzzy Grouping Particle Swarm Optimization for Care Task Assignment

10.1 INTRODUCTION

Healthcare task assignment in a hospital setting involves allocation of nursing care activities to nursing staff on a daily basis, subject to hard and soft constraints regarding relationships between tasks and capacity limitation of nurses (Paulussen et al., 2003; Vermeulen et al., 2006). The responsibility of nurses is to efficiently and effectively provide high-quality care to patients. However, due to a worldwide shortage of nurses and the ever-increasing pressure for high-quality nursing care, care task assignment is a crucial but difficult problem.

In most hospitals, task assignment is done manually using spreadsheets based on predetermined patient information regarding execution time intervals in which care activities should be done to patients (Cheng et al., 2008). Normally, tasks are assigned to nurses according to basic rules of thumb (Cheng et al., 2007). However, this procedure may yield poor and unfair task schedules, leading to poor quality of service. Developing efficient and effective task scheduling methods is imperative.

High-quality schedules should satisfy, as much as possible, the preferences and expectations of the patients, the nursing staff, and management. Care task schedules should ensure that the actual care execution times are as close as possible to the desired time windows prespecified by patients. This leads to high patient satisfaction. In addition, care tasks should be assigned fairly among the available nurses; the goal is to balance, as much as possible, the individual workloads assigned to nurses. This ultimately leads to high worker morale, service efficiency, and job satisfaction. However, in satisfying the preferences of patients and nurses, the decision maker should consider management goals and expectations (Mutingi and Mbohwa, 2014a,b). However, management goals are often qualitative and imprecise, adding to the complexity of the problem.

In the presence of imprecise and conflicting preferences and management goals, the use of conventional optimization methods, such as linear programming, and basic dispatching heuristics, such as earliest due date, slack, and first in/first out, is limited (Cheng et al., 2008; Mutingi and Mbohwa, 2012, 2014a). For instance, conventional dispatching rules disallow the use of multiple criteria in the scheduling process. Moreover, the rules have a rigid structure that excludes the use of other useful information that may be available. Thus, the care task assignment problem is characterized by complicating features:

1. The presence of fuzzy staff preferences and wishes, such as fairness and equity on assigned workloads
2. The presence of fuzzy patient expectations and preferences on time windows and care due dates
3. The presence of imprecise management goals and choices that are difficult to quantify
4. The need to find a fair trade-off between conflicting goals of the problem

Designing interactive metaheuristics to handle fuzzy goals and preferences is imperative. This will provide high-quality task schedules that eventually lead to improved care worker satisfaction (job satisfaction), service efficiency, service quality, and business competitiveness. Incorporating fuzzy evaluation techniques into metaheuristic approaches is a viable and promising option. The purpose of this research is to develop a fuzzy heuristic approach to care task assignment in a hospital setting from a multicriteria perspective. Therefore, the specific research objectives are as follows:

1. To describe the care task assignment problem
2. To propose a fuzzy grouping particle swarm optimization
3. To provide illustrative examples, demonstrating the effectiveness of the algorithm

The next section presents a brief description of the care task assignment. This is followed by a description of the proposed fuzzy grouping particle swarm optimization algorithm. Computational experiments, results, and discussions are presented. The chapter ends with a summary of the research work covered.

10.2   THE CARE TASK ASSIGNMENT PROBLEM

This section provides a description of the healthcare task assignment problem and its formulation.

10.2.1   PROBLEM DESCRIPTION

The healthcare task assignment problem (CTAP) is concerned with the allocation of a set of healthcare tasks to nursing staff so that patients can receive the required healthcare service (Aiken, Clarke, and Sloane, 2002; Bard and Purnomo, 2005; Mutingi and Mbohwa, 2014a). The essence of the problem is that all the tasks must be assigned, subject to a set of constraints concerning caregiver capacity, the nature of tasks, and their precedence relationships. In most hospitals, care tasks are assigned based on predetermined patient information on task duration and the time window during which specific tasks should be performed. The task assignment process is often carried out manually using spreadsheets. In practice, standard work

Date	03 Jan 2015								
Time	0800	0930	1000	1030	1100	1130	1200	1230	1300
Patient 1	Shampoo toilet assist bath	Check gauze		BP check			BP check	Pre-prandial medicine	Serve, clear tray clean teeth	...
Patient 2	BP check	Antibiotic		BP check instil drops			Check compliance	Serve, clear tray clean teeth		...
	:	:	:	:	:	:	:	:	:	.

FIGURE 10.1   A typical example of a daily worksheet.

procedures are recorded in manuals that contain key information on care tasks and the associated resources needed. Care tasks—for instance, assistance with meals, instillation of drops, and preparation of infusions—can be categorized into preparatory tasks, executions tasks, and cleanup tasks (Cheng et al., 2007). Therefore, each care activity consists of these three types of tasks. Figure 10.1 shows an example of a daily worksheet.

CTAP is analogous to the job dispatching or job shop scheduling problem. Similarly to job dispatching, CTAP can be addressed using dispatching rule-based methods. In that context, appropriate priority rules, such as first in/first out (FIFO) and earliest due date (EDD), can be applied. Deriving from this analogy, we identified a number of constraints that were classified into hard constraints and soft constraints.

10.2.2 PROBLEM FORMULATION

Table 10.1 presents the constraints associated with the CTAP problem (Cheng et al., 2007). Hard constraints must always be satisfied, while soft constraints may be violated, but at a penalty cost. Hard constraints are concerned with task release time, task precedence, and staff capacity. For clarity of deliberation on the CTAP problem formulation, we define the following notations:

u	Index for U activities, $u = 1,2,...,U$
v	Index for V_u tasks in activity u, $v = 1,2,...,V$
d_u	The desirable due date of activity u
r_u	The release time of activity u
t_{uv}	Starting time for care task v of activity u
$tc_{v1,v2}$	The changeover time from task $v1$ to $v2$
p_{uv}	The expected processing time of task v of activity u

10.2.2.1 Release Time Constraints

The release time of a task pertains to the earliest time when that particular task is ready for execution.

$$t_{uv} \geq r_u \quad \forall u, u = 1,2,...,U \quad \forall v, v = 1,2,...,V_u \qquad (10.1)$$

TABLE 10.1
Typical Care Task Constraints

Constraint	Brief Description
	Hard Constraints
1. Release time constraints	The first nursing task of a certain nursing activity cannot be handled until the release time of the activity.
2. Precedence constraints	A nursing task cannot be handled until the previous task of the same nursing activity is finished.
3. Capacity constraints	Nurses have limited processing capacity to handle their work. In general, they can handle only one nursing task at a time.
	Soft Constraints
1. Due date constraints	The last nursing tasks of a certain nursing activity should be finished before the due date of the nursing activity.
2. Transition time constraints	Some nursing tasks cannot be handled immediately after the previous task of the same nursing activity is finished.
3. Time window constraints	Some execution tasks of nursing activities should be handled within an expected execution time interval.

10.2.2.2   Precedence Constraints

Precedence constraints relate to the structural sequence of tasks that should be observed when executing specific tasks. It follows that following care tasks cannot be handled until their predecessors are finished.

$$t_{u,v-1} + p_{u,v-1} \leq t_{uv} \quad \forall u, u = 1, 2, ..., U \quad \forall v, v = 1, 2, ..., V_u \quad (10.2)$$

10.2.2.3   Capacity Constraints

Capacity constraints limit the number of tasks that a nurse can perform at any given time. In this case, we assume that a nurse can perform only one task at a time. Therefore,

$$t_{u1,v1} + p_{u1,v1} + tc_{v1,v2} \leq t_{u2,v2} \quad (10.3)$$

where $v1$ and $v2$ represent any two tasks from any activities to be performed by a specific nurse.

In addition to the preceding three hard constraints, soft constraints should be satisfied as much as possible. We identify three types of soft constraints concerned with due date, task changeover, and execution time (time window) of the tasks.

10.2.2.4   Due-Date Constraints

The due-date constraints ensure that the end time of each activity is as close as possible to the desirable due date of that activity. This restriction can be represented by the following expression:

$$t_{uv} + p_{uv} \leq d_u \quad \forall u, u = 1, 2, ..., U \quad \forall v, v = 1, 2, ..., V_u \quad (10.4)$$

10.2.2.5   Transitions Time Constraints

These constraints ensure that the changeover time between successive tasks is as close to the desired time as possible. This implies that succeeding tasks should not be performed immediately after their predecessors, but rather after a desired lapse of time (or transition time) $trans_{v1v2}$ is reached.

$$t_{uv2} \approx t_{uv1} + p_{uv1} + trans_{v1,v2} \quad \forall u, \forall v1, \forall v2 \quad (10.5)$$

where $v1$ and $v2$ are tasks of the same activity u, and task $v2$ can only start after task $v1$ is completed, with a transition time $trans_{v1,v2}$.

10.2.2.6   Time Window Constraints

Time window constraints limit, as much as possible, the execution time of some care activities to be within the desired time window $\left[T_u^1, T_u^2 \right]$, where T_u^1 and T_u^2 are the lower and upper bounds on the expected execution time of an activity u, respectively. For instance, lunch meals may be restricted to time window [1200, 1300] hours. Therefore,

$$t_{uv} \geq T_u^1 \text{ and } t_{uv} + p_{uv} \leq T_u^2, \quad \forall u, \forall v \quad (10.6)$$

where T_u^1 and T_u^2 are the lower and upper bounds on the execution time of activity u, respectively.

10.2.3   Problem Objectives

The objectives of the CTAP are (1) to maximize fairness in workload assignment and (2) to minimize violation of soft constraints. Thus, the aim is to maximize the quality of the care schedule by finding a trade-off between these objectives. Clearly, the CTAP is a complex problem that is difficult to solve using conventional solution approaches. To this end, we present an enhanced fuzzy grouping particle swarm optimization algorithm for interactive decision making for the problem.

10.3   A FUZZY GROUPING PARTICLE SWARM OPTIMIZATION

The fuzzy grouping particle swarm optimization (FGPSO) algorithm is a development from particle swarm optimization. It uses fuzzy theoretic techniques to

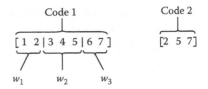

FIGURE 10.2 FGPSO coding scheme.

evaluate the performance of alternative solutions. FGPSO takes advantage of the group structure of the problem. The algorithm and its elements, including particle position representation, initialization, fuzzy fitness evaluation, and velocity and position update are presented in this section.

10.3.1 FGPSO CODING SCHEME

To enhance the performance of FGPSO, a unique group coding scheme is developed to exploit the group structure of the problem (see Figure 10.2). Let $C = \{1, 2, 3,...,V\}$ be a particle representing a set of V tasks to be performed by I nurses. Then, the evaluation of C involves partitioning tasks along C into g groups such that all the hard constraints are satisfied and the violation of soft constraints is minimized. For instance, given seven tasks ($V = 7$) and three nursing staff ($I = 3$), the group structure of the problem is coded as shown in Figure 10.2. The structure consists of two codes: code 1 represents the assignment of care workers w_1, w_2, and w_3 to groups of tasks {1,2}, {3,4,5}, and {6,7}, respectively. Genetic operators work on code 1, while code 2 records the position of the delimiter or frontier "|," which separates task groups.

10.3.2 INITIALIZATION

An initial population of candidate solutions is created by randomly assigning tasks to nurses. First, care tasks are arranged in ascending order of their expected start times. For each nurse, unassigned tasks are probabilistically allocated, beginning with the earliest. By this procedure, the algorithm increases the likelihood of generating a good initial population of feasible solution candidates.

10.3.3 FITNESS EVALUATION

The fitness evaluation procedure determines the fitness of each candidate solution based on a combination of fuzzy evaluation functions. The functions should measure the relevant quality of the candidate solution and capture the imprecise conflicting goals and constraints. The evaluation function, F_t, at iteration t should be a normalized function obtained from its n constituent normalized functions denoted by μ_h ($h = 1,...,n$). Therefore, we use a fuzzy multifactor evaluation method:

$$F_t(s) = \sum_h w_h \mu_h(s)$$ (10.7)

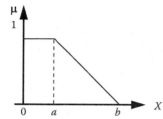

FIGURE 10.3 Interval-valued linear membership function.

where s is a candidate solution at iteration t, and w_h denotes the weight of the function μ_h. The use of the max–min operator is avoided so as to prevent possible loss of vital information.

The fitness or quality of a solution is a function of how much it satisfies the preferences of nurses and patients, as well as management goals and choices. Due to ambiguity and imprecision of these preferences and goals, fitness is modeled as a normalized interval-valued fuzzy membership function, as shown in Figure 10.3.

The satisfaction level is represented by a decreasing linear function where $[0,a]$ is the most desirable range, and b is the maximum acceptable. Therefore, the corresponding function is as follows:

$$\mu_A(x) = \begin{cases} 1 & \text{if } x \le a \\ (b-x)/(b-a) & \text{if } a \le x \le b \\ 0 & \text{if otherwise} \end{cases} \tag{10.8}$$

10.3.3.1 Membership Function 1: Due Date

The completion time of every activity should, as much as possible, be within acceptable limits so as to maximize the quality of the care schedule. Let t_u and d_u be the completion time and due date of each activity u, respectively. Then the total variation of completion times from their respective due dates is given by the expression

$$z_1 = \sum_u |t_u - d_u| \tag{10.9}$$

Assuming that the interval-valued membership function holds, we obtain the following expression:

$$\mu_1 = \mu_A(z_1) \tag{10.10}$$

10.3.3.2   Membership Function 2: Transition Time

For any activity u, assume that task $v2$ is supposed to start soon after task $v1$, with a transition time of $trans_{v1,v2}$. Then, the objective is to minimize the total variation, z_2, from meeting this transition requirement, represented by

$$z_2 = \sum_u \sum_{v1,v2 \in J_u} \left| t_{u,v1} + p_{u,v1} + trans_{v1,v2} - t_{u,v2} \right| \qquad (10.11)$$

Therefore, assuming the interval-valued membership function, we obtain

$$\mu_2 = \mu_A(z_2) \qquad (10.12)$$

10.3.3.3   Membership Function 3: Time Windows—Earliness

The fitness of a candidate care schedule can be measured in terms of total earliness. Let the total earliness be denoted by z_3. Then, it follows that

$$z_3 = \sum_u \sum_{v \in J_u} \max\left(0, T_u^1 - t_{uv}\right) \qquad (10.13)$$

Here, J_u is a set of tasks from activity u, and T_u^1 is the lower bound of the time window of activity u. Assuming that e follows a trapezoidal linear membership, the fuzzy membership function for earliness can be expressed as follows:

$$\mu_1 = \mu_A(z_3) \qquad (10.14)$$

10.3.3.4   Membership Function 4: Time Window—Lateness

Apart from earliness, the fitness of a candidate care schedule can also be measured in terms of total lateness or tardiness. Therefore, the objective is to minimize total tardiness, z_4, given by the expression

$$z_4 = \sum_u \sum_{v \in J_u} \max\left(0, t_{uv} + p_{uv} - T_u^2\right) \qquad (10.15)$$

Here, T_u^2 is the upper bound on time window of activity u, and J_u is a set of tasks in activity u. Assume that z_4 follows a trapezoidal linear membership. Then the fuzzy membership function for tardiness can be expressed as follows:

$$\mu_4 = \mu_4(z_4) \qquad (10.16)$$

10.3.3.5 Membership Function 5: Workload Fairness

Let h_i denote the workload of nurse i, and a be the average workload. The objective is to minimize workload variation z_5 as given by the following:

$$z_5 = \sum_i |h_i - a| \tag{10.17}$$

Here, the workload h_i is given by the expression

$$h_i = \sum_u \sum_v p_v x_{iv} \qquad \forall i, i = 1, 2, \ldots, N \tag{10.18}$$

where x_{iv} is a binary variable representing whether or not task v is assigned to nurse i. We use a fuzzy membership function μ_5 of the form

$$\mu_5 = \mu(z_5) \tag{10.19}$$

10.3.4 VELOCITY AND POSITION UPDATE

As in the PSO mechanism, FGPSO makes use of a velocity vector to update the current position of each particle in the swarm. The velocity and the position updates of each particle at iteration $(t + 1)$ are determined by the following expressions, respectively:

$$v_i(t + 1) = w \cdot v_i(t) + c_1 \cdot \eta_1 \cdot (pbest_i(t) - x_i(t)) + c_2 \cdot \eta_2 \cdot (gbest(t) - x_i(t)) \tag{10.20}$$

$$x_i(t + 1) = x_i(t) + v_i(t + 1) \tag{10.21}$$

where
w = an inertia weight showing the effect of previous velocity on the new velocity vector
c_1 and c_2 = constants
η_1 and η_2 = uniformly distributed random variables in the range [0,1]

10.4 THE OVERALL FGPSO ALGORITHM

The previous sections described the procedures of operators of the FGPSO algorithm. Deriving from these procedures, the overall structure of the FGPSO algorithm is presented in the form of a pseudocode in Figure 10.4.

The algorithm accepts input of the parameters w, η_1, η_2, c_1, c_2, and N from the user. A population of N candidate solutions is generated at random. Depending on

Algorithm 1. Fuzzy grouping particle swarm optimization algorithm
1.   Input w, η_1, η_2, c_1, c_2, N;
2.   Initialization;
3.   **For** $i = 1$ to N;
4.   Initialize particle position $x_i(0)$ and velocity $v_i(0)$;
5.   Initialize $pbest_i(0)$;
6.   **End For**;
7.   Initialize $gbest(0)$;
8.   **For** $i = 1$ to N:
9.   Compute fuzzy fitness $f(x)$, $x = (x_1, x_2,...,x_N)$;
10. **Repeat**
11.   **For** $i = 1$ to N:
12.   Compute fuzzy fitness f_i;
13.   **If** $(f_i >$ current $pbest)$ **Then**
14.   Set current value as new $pbest$;
15.   **If** $(f_i >$ current $gbest)$ **Then**
16. $gbest = i$;
17.   **End If**;
18.   **End For**;
19.   **For** $i = 1$ to N:
20.   Find neighborhood best;
21.   Compute particle velocity $v_i(t + 1)$;
22.   Update particle position $x_i(t + 1)$;
23.   **End For**;
24. **Until** (Termination criteria are satisfied);

FIGURE 10.4   A pseudocode for the FGPSO algorithm.

the nature of the variables of the problem, real- or integer-valued or candidate solutions are generated based on the group coding scheme.

The proposed FGPSO algorithm has a number of desirable advantages in its application. First, its procedure is easy to follow and therefore can be implemented easily in a number of problem situations. It is robust and versatile in that it is applicable to similar problems with little or no fine-tuning. Moreover, the algorithm is computationally efficient, obtaining good solutions within reasonable computation times. Most importantly, incorporating fuzzy evaluation into the algorithm allows the global optimization process to pass through inferior solutions, which will eventually yield improved solutions. Fuzzy multicriteria evaluation ensures that instances of infeasible solutions are avoided during algorithm execution.

10.5 COMPUTATIONAL EXPERIMENTS AND RESULTS

To test the proposed algorithm, three sets of problems were used; we randomly generated the data for the care task assignment problem.

10.5.1 COMPUTATIONAL EXPERIMENTS

In generating test problems, assume a hospital setting where a group of nurses work in a day shift from 8:00 a.m. to 5:00 p.m. Further, assume a normal distribution for the processing times, p_{uv}, is generated according to the expression, $p_{uv} \in [\overline{p}_{uv} - \varepsilon \cdot \overline{p}_{uv}, \overline{p}_{uv} + \varepsilon \cdot \overline{p}_{uv}]$, where $\varepsilon \in [0,1]$. As such, problems were randomly generated, with different processing times by setting $\varepsilon = 0.5$. The release times r_u for each activity u, time windows for specific tasks, and the due dates for different activities were also created randomly between 8:00 a.m. and 5:00 p.m.

The algorithm was coded in Java 7™, standard edition, a Windows 7 operating system on a PC running on an Intel Pentium 3.0 GHz and 4GB RAM. The performance of FGPSO was compared against particle swarm optimization (PSO) and the genetic algorithm (GA), which were developed in this study (Mutingi and Mbohwa, 2014a).

10.5.2 Results and Discussions

10.5.2.1 Experiment 1

Consider a hypothetical problem consisting of five nurses and 30 tasks, with a known optimal solution, $F_t = 1$. The aim of the experiment is to test the consistency, reliability, and effectiveness of each algorithm when in use. It follows that the higher the success rate is, the higher the effectiveness will be, and vice versa. Table 10.2 shows the results of the experiment. All the algorithms obtained the desired optimal solution. In terms of search success rate, FGPSO rated 100%, followed by PSO, rated 93.33%; the GA rated 90%. Regarding computational efficiency, FGPSO performed the best with CPU time of 4.8 seconds. On the other hand, the average CPU times for PSO and the GA were 5.5 and 6.2 seconds, respectively. Therefore, FGPSO has a higher potential for efficient and effective performance, even on larger scale problems.

Figure 10.5 shows a transcription of the fitness values over 500 iterations. As can be seen from the graphical analysis, the final fitness value of FGPSO is higher than basic PSO, which in turn is higher than the basic GA. Moreover, the FGPSO run approached unity much faster than the other algorithms. The comparative performance of FGPSO against basic PSO and the GA shows that FGPSO outperforms the two competitive algorithms in terms of efficiency and the final fitness value.

10.5.2.2 Experiment 2

The algorithms were tested for search success rate based on a hypothetical problem consisting of four nurses and 24 tasks, with a known optimal solution. Each algorithm was run 30 times, recording the average computation (CPU) time and the

TABLE 10.2

Comparative Analysis

Algorithm	Average CPU Time (sec)	Search Success Rate	Solution, F_t
FGPSO	4.8	100.0%	1.00
PSO	5.5	93.3%	1.00
GA	6.2	90.0%	1.00

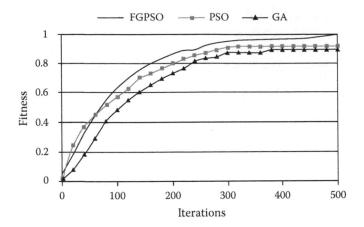

FIGURE 10.5 Comparative performance of FGPSO, PSO, and GA.

search success rate. Table 10.3 shows the results of the experiment. All the algorithms obtained the desired optimal solution. In terms of search success rate, FGPSO rated 100%, followed by PSO, rated 93.3%, while the GA rated 87.7%. Regarding computational efficiency, FGPSO performed the best with a CPU time of 4.8 seconds. On the other hand, the average CPU times for PSO and the GA were 5.7 seconds and 7.1 seconds, respectively. Therefore, FGPSO has a higher potential for efficient and effective performance, even on larger scale problems.

10.5.2.3 Further Experiments

In this experiment, results for 15 randomly generated problems are reported. To compare the potential efficiency and effectiveness of the algorithm, the computational time (CPU time) was used as a measure of performance. For each problem case, 30 independent runs were executed. The stopping criteria were determined by the number of iterations without solution improvement, $T_I < 30$, with a maximum number of iterations set at $T_m = 300$.

Table 10.4 presents a summary of the comparative analysis of the performance of FGPSO in terms of average CPU times. FGPSO is compared with the basic GA and PSO. Though the basic GA and the PSO algorithm may be fairly competitive over small-scale problems such as problems 1 and 2, they performed below the FGPSO over the reset of the larger problems. Overall, the mean computation time for FGPSO

TABLE 10.3
A Comparative Analysis of FGPSO and Other Algorithms

Algorithm	Average CPU Time (sec)	Search Success Rate	Solution, F_t
FGPSO	4.8	100.0%	1.00
PSO	5.7	93.3%	1.00
GA	7.1	87.7%	1.00

TABLE 10.4

Comparison Analysis between FGPSO and Other Algorithms

Problem	No. of Tasks	No. of Nurses	Average CPU Time (sec)		
			Basic GA	Basic PSO	FGPSO
1	24	5	4.2	4.3	3.2
2	35	6	3.8	4.2	4.8
3	40	8	4.9	4.6	4.8
4	45	8	7.2	8.1	7.3
5	50	9	10.6	9.6	7.9
6	55	9	11.4	13.3	10.3
7	60	10	19.7	18.4	10.9
8	65	10	21.1	17.3	11.2
9	70	11	22.3	19.9	16.3
10	75	11	29.6	27.2	22.4
11	90	12	37.4	36.9	32.1
12	95	12	49.9	39.5	33.6
13	105	13	52.6	48.6	37.1
14	120	16	51.6	51.3	45.1
15	130	20	69.5	66.8	68.7
	Mean	18	31.13	28.88	21.05

was 21.05 seconds, compared to the basic GA with 28.88 seconds, and basic PSO with 21.06 seconds.

From the preceding comparative analyses, FGPSO is efficient and effective. It is capable of producing satisfactory solutions within reasonable computations times, even over larger problems.

10.6   SUMMARY

Care task assignment is a common problem in hospitals; it is concerned with finding the best way to allocate care tasks to a limited pool of nurses so that all tasks are performed in as timely a manner as possible, nurse workloads are assigned fairly, transition times between tasks are satisfied as much as possible, and expert choices are taken into account. The higher the satisfaction level of these requirements is, the higher the quality of the task schedule will be. This is a hard problem that demands interactive fuzzy heuristic methods. This chapter presented a fuzzy grouping particle swarm algorithm to solve the problem. By exploiting permutations of groups of tasks across candidate task schedules and within each candidate schedule, and using enhanced heuristic operators, the algorithm can address the problem efficiently. The approach provides useful contributions to researchers and practitioners in healthcare.

The proposed algorithm contributes to knowledge in flexible, adaptable, and interactive heuristic optimization methods. By realizing the need to holistically satisfy the patient, healthcare worker, and management, this research provides a judicious trade-off approach by which the three players in a healthcare system can be satisfied, with potential long-term benefits. Moreover, the problem can be modeled with more realism, considering fuzzy expert choices of the decision maker. The method presents built-in heuristic techniques that exploit the group structure of the problem to handle large-scale problems efficiently. Thus, the proposed algorithm is an invaluable addition to the body of knowledge in healthcare operations management.

The practicing decision maker can benefit from the suggested approach to the CTAP in a number of ways. The algorithm provides an opportunity to use weights to interactively incorporate the decision maker's preferences and choices. In practice, decision makers appreciate the use of an interactive decision support that provides a list of good alternative solutions from which the most appropriate decision can be chosen, taking into account other practical considerations. In this view, expert knowledge can be incorporated into the decision process, unlike when prescriptive optimization methods are used. Overall, the proposed algorithm is a viable tool in care task assignment.

REFERENCES

Aiken, L. H., Clarke, S. P. and Sloane, D. M. 2002. Hospital staffing, organization, and quality of care: Cross-national findings. *International Journal for Quality in Health Care* 14(1): 5–13.

Bard, J. F. and Purnomo, H. W. 2005. Real-time scheduling for nurses in response to demand fluctuations and personnel shortages. *Proceedings of the 5th International Conference on Scheduling: Theory & Applications (MISTA'05)*, 397–406.

Cheng, M., Ozaku, H. I., Kuwahara, N., Kogure, K. and Ota, J. 2007. Nursing care scheduling problem: Analysis of staffing levels. *Proceedings of the 2007 IEEE International Conference on Robotics and Biomimetics*, Sanya, China, December 15–18.

Cheng, M., Ozaku, H. I., Kuwahara, N., Kogure, K. and Ota, J. 2008. Simulated annealing algorithm for scheduling problem in daily nursing cares. *Proceedings of the IEEE International Conference on Systems, Man and Cybernetics*, SMC, 1681–1687, October.

Mutingi, M. and Mbohwa, C. 2012. Enhanced group genetic algorithm for the heterogeneous fixed fleet vehicle routing problem. *IEEE IEEM Conference on Industrial Engineering and Engineering Management*, Hong Kong, 207–211, December.

Mutingi, M. and Mbohwa, C. 2014a. Care task assignment: A simulated metamorphosis approach (working paper), unpublished.

Mutingi, M. and Mbohwa, C. 2014b. Home healthcare staff scheduling: A clustering particle swarm optimization approach. *Proceedings of the 2014 International Conference on Industrial Engineering and Operations Management*, Indonesia, 303–312, January 7–9.

Paulussen, T. O., Jennings, N. R., Decker, K. S. and Heinzl, A. 2003. Distributed patient scheduling in hospitals. *Proceedings of the 18th International Joint Conference on Artificial Intelligence*, 1224–1229.

Vermeulen, I., Bohte, S., Somefun, K. and Poutre, H. L. 2006. Improving patient activity schedules by multi-agent Pareto appointment exchanging. *Proceedings of the 8th IEEE Conference on E-Commerce Technology (CEC'06)*.

11 Future Trends and Research Prospects in Healthcare Operations

11.1 SUMMARY

Decision making in staff scheduling in healthcare organizations is influenced by imprecise goals, preferences, and expectations of (1) patients, (2) management, and (3) workers. As such, the decision-making process often takes place under fuzzy goals and constraints. Where management goals, staff preferences, patients' preferences, and the consequences of the decisions taken are not precisely known at the planning stage, staff scheduling is complex. Motivated by the presence of imprecise, conflicting goals and preferences in healthcare service operations, the theme of this book was about developing fuzzy approaches to staff scheduling problems.

The overall purpose of this book was to develop fuzzy interactive multicriteria approaches that address the inherent uncertainties in staff scheduling problems. This was reflected in the research objectives of the book:

1. To analyze staff scheduling problems, from a fuzzy multicriteria viewpoint, highlighting their potential limitations
2. To develop interactive fuzzy metaheuristic approaches to staff scheduling, considering fuzzy conflicting goals and preferences
3. To develop an interactive fuzzy metaheuristic approach for homecare staff scheduling, considering fuzzy conflicting goals and preferences
4. To develop an interactive fuzzy metaheuristic for care task assignment in a hospital setting with fuzzy conflicting goals and preferences

The purpose of this chapter is to highlight the research achievements and contributions arising from these objectives. Finally, further research prospects are presented.

Healthcare staff scheduling is complex but commonplace in various healthcare settings across the globe. Decision making is a challenge to most healthcare managers due to lack of availability of appropriate and effective decision tools. As such, any advances in the development of efficient, effective, and interactive scheduling approaches are highly significant.

This book focused on research on healthcare staff scheduling and the application of modern fuzzy heuristic optimization approaches. Healthcare operations, in both hospital and home healthcare settings, are inundated with complex fuzzy features that impose difficulties in the construction of work schedules. Staff scheduling is impacted by imprecise or vague management goals, staff preferences, and patients' preferences. In the presence of fuzzy variables, the impact of the decisions taken is unknown at the planning stage. Consequently, the decision maker has to rely on the use of expert intuition and knowledge. This research presented fuzzy optimization approaches that incorporate the concepts of fuzzy sets and metaheuristic algorithms for staff scheduling in healthcare organizations.

The book comprised three sections. The first section highlighted recent research trends and challenges in healthcare staff scheduling, necessitating the use of fuzzy theory concepts and fuzzy evaluation techniques. The second section presented modern metaheuristic approaches and the use of fuzzy set theory concepts to address complex multiobjective decisions. A novel fuzzy simulated metamorphosis (FSM) algorithm and an enhanced fuzzy simulated evolution (FSE) algorithm were proposed. Furthermore, a unique fuzzy grouping genetic algorithm (FGGA) and an enhanced fuzzy grouping particle swarm optimization (FGPSO) algorithm were presented. The third section focused on research applications in healthcare staff scheduling, providing researchers and practitioners a practical, in-depth understanding of fuzzy metaheuristic approaches. To model the fuzzy features of staff scheduling problems, conflicting fuzzy management goals, staff preferences, and patient expectations were expressed and evaluated as fuzzy membership functions.

The research applications of the approaches fall into two categories. The first is concerned with constructing high-quality staff schedules spanning over a period of a week or more. The second category focuses on developing high-quality staff schedules in a home healthcare setting. The overall objective is to optimize costs, violation of staff and patient preferences, subject to time and staff capacity constraints.

The emerging approaches proposed in this book are beneficial in three main ways: (1) they incorporate fuzzy preferences and decision makers' choices, giving more realism to the approaches; (2) they are flexible and adaptive to problem situations, providing room for interactive decision support for decision makers; and (3) they provide reliable solutions within reasonable computation times. Therefore, these approaches are highly useful to researchers, academicians, and practicing decision makers concerned with healthcare staff scheduling.

11.2   RESEARCH CONTRIBUTIONS

This research presented fuzzy multicriteria heuristic approaches, incorporating fuzzy evaluation techniques and metaheuristics in an interactive manner. The research achievements contribute to knowledge development and practice in healthcare staff scheduling.

11.2.1   Contributions to Knowledge

The study is a significant contribution to knowledge in healthcare staff scheduling. In the first place, the critical review of various scheduling methods is invaluable to academicians and researchers. Second, the study proposes a unique holistic view of scheduling where preferences and choices of the patient, the healthcare worker, and management are simultaneously considered, so that all three players in a healthcare system are satisfied. The proposed fuzzy multicriteria evaluation offers a judicious way of modeling complex scheduling problems with conflicting objectives of varying weight or importance. Third, the study emphasizes the development of interactive approaches that provide a pool of good alternative solutions, rather than prescribe a single solution to the decision maker. These realizations contribute to further knowledge in healthcare staff scheduling.

11.2.2 CONTRIBUTIONS TO PRACTICE

This study has demonstrated that fuzzy multicriteria heuristic approaches can efficiently and effectively model and solve healthcare staff scheduling problems where fuzzy goals and preferences should be satisfied as much as possible. The main practical-oriented contributions of the research are fourfold:

1. The study presented fuzzy set theory based methods that provide effective ways to capture and model fuzzy choices, preferences, and expectations, which are difficult to precisely quantify in practice. Implementing fuzzy evaluation techniques makes fitness evaluation of candidate solutions feasible.
2. The suggested fuzzy multicriteria approaches provide the decision maker an opportunity to incorporate managerial choices and preferences in the form of weights, which is helpful for developing solutions that satisfy patients, management, and healthcare staff.
3. The suggested fuzzy multicriteria algorithms can judiciously provide practical and effective staff scheduling decisions in the presence of fuzzy conflicting objectives.
4. The interactive fuzzy heuristic approaches can serve as better decision support tools by providing a set of good alternative solutions, contrary to conventional methods that prescribe a single solution.
5. The proposed fuzzy algorithms provide near-optimal solutions within reasonable computation times, which is important to practicing decision makers where effective decisions are needed within limited time frames.

Overall, the study opens up knowledge avenues for researchers, academicians, and practitioners in the operations research and operation management community. However, the study has faced some limitations.

11.3 FURTHER RESEARCH CONSIDERATIONS

With the growing interest in healthcare operations management, a number of potential areas of research can be realized. Some of the potentially active study areas are bed allocation, patient scheduling, and operating room scheduling. These prospective research areas are explained in the following sections.

11.3.1 BED ALLOCATION

The hospital bed is one of the most crucial resources of the medical facility (Green and Nguyen, 2001; Gorunescu, McClean, and Millard, 2002). A low supply of hospital beds may lead to low patient admission, whereas surplus hospital beds may lead to resource wastage (Kao and Tung, 1981). Moreover, a typical hospital comprises a number of wards that serve patients with different healthcare service needs. Allocating beds among these wards is limited by the availability of resources (Gorunescu, McClean, and Millard, 2002; Green, 2006; Gong, Zhang, and Zhun Fan, 2010).

Estimating the requirement of bed allocation a priori is not trivial since this is often characterized with uncertainties. As a result, beds may be overloaded in one ward, while underloaded in the other. Therefore, efficient and effective methods are essential for bed allocation so as to improve the overall service level and resource usage.

A significant number of methods have been suggested to solve the bed allocation problem, including expert systems, heuristics, and queuing theory (Green and Nguyen, 2001; Nguyen et al., 2005; Green, 2006; Li et al., 2009). Common objectives in solving this problem are as follows:

1. To maximize the patient admission rate
2. To maximize the overall bed occupancy
3. To simultaneously satisfy the objective in (1) and (2)

However, past methods rarely consider uncertainties in the problem. It will be interesting to consider the inherent imprecise parameters and model using fuzzy-based methodologies, preferably from a multicriteria decision perspective.

11.3.2 PATIENT SCHEDULING

Often treated as a job shop problem, patient scheduling requires that each patient be booked for possibly multiple appointments, which should follow a prespecified multiday pattern (Wang, Ma, and Guan, 2009; Kanagaa and Valarmathi, 2012). In some cases, each appointment may need several care activities, which also follow a prespecified intraday pattern. In general, the main objectives are as follows:

1. To satisfy the preferences of patients
2. To minimize patients' waiting times
3. To balance workload, minimizing the peaks of nurses' workloads in an outpatient setting

Though the patient scheduling problem has received significant attention, most of the extant studies have addressed the problem using crisp approaches (Wang, Ma, and Guan, 2009). However, when dealing with such problems, features such as processing times, release dates, due dates, management goals, and patient preferences are often uncertain. Therefore, in such situations, fuzzy-based approaches are most applicable to capture and model uncertainties using fuzzy numbers instead of estimated crisp numbers (Tang, Yan, and Cao, 2014).

In view of this, the patient scheduling problem is a potential study area to consider applying fuzzy heuristic methods.

11.3.3 OPERATING ROOM SCHEDULING

Surgical sector expenses account for more than 33% of the projected hospital budget due to expensive specialized labor, such as surgeons and anesthetists, and material costs, such as beds and surgical equipment (Fei, Meskens, and Chu, 2010; Meskens, Duvivier, and Hanset, 2013). Recent studies show that hospital managers desire to find cost-effective ways of running the operating room in order to improve the quality of service.

In general, proper surgical planning and scheduling approaches will assist hospital managers to utilize operating rooms as efficiently as possible (Augusto, Xie, and Perdomo, 2010). Common objectives when making decisions in this area are as follows:

1. To satisfy the healthcare service needs of patients as much as possible
2. To minimize the overall healthcare operating costs
3. To satisfy the preferences of the surgeons, considering the human and material resource constraints

Operating room management consists of planning and scheduling of surgical cases (Guinet and Chaabane, 2003). While planning focuses on the date of surgery, daily scheduling looks at the sequence of operations in each room each day. These problems are subject to various uncertainties. The operating room scheduling problem is often characterized with long waiting times, uncertain cancelations, and resource overload (Jebali, Alouane, and Ladet, 2006).

The preceding research areas and directions can help to advance the knowledge and practice of fuzzy heuristic approaches to address the uncertain features in healthcare operations management problems.

REFERENCES

Augusto, V., Xie, X. and Perdomo, V. 2010. Operating theatre scheduling with limited recovery beds and patient recovery in operating rooms. *Computers and Industrial Engineering* 58: 231–238.

Fei, H., Meskens, N. and Chu, C. 2010. A planning and scheduling problem for an operating theatre using an open scheduling strategy. *Computers and Industrial Engineering* 58: 221–230.

Gong, Y.-J., Zhang, J. and Zhun Fan, Z. 2010. A multiobjective comprehensive learning particle swarm optimization with a binary search-based representation scheme for bed allocation problem in general hospital. *The 2010 IEEE International Conference on Systems Man and Cybernetics* (SMC), 1083–1088.

Gorunescu, F., McClean, S. I. and Millard, P. H. 2002. Using a queueing model to help plan bed allocation in a department of geriatric medicine. *Health Care Management Science* 5 (4).

Green, L. 2006. Queuing analysis in healthcare. In *Patient flow: Reducing delay in healthcare delivery*, ed. R. Hall, 281–308, Springer, New York.

Green, L. V. and Nguyen, V. 2001. Strategies for cutting hospital beds: The impact on patient service. *Health Services Research* 36 (2): 421–442.

Guinet, A. and Chaabane, S. 2003. Operating theatre planning. *International Journal of Production Economics* 85: 69–81.

Jebali, A., Alouane, A. B. H. and Ladet, P. 2006. Operating rooms scheduling. *International Journal of Production Economics* 99: 52–62.

Kanagaa, E. G. M. and Valarmathi, M. L. 2012. International Conference on Communication Technology and System Design 2011, Procedia Engineering 30: 386–393.

Kao, E. P. C. and Tung, G. G. 1981. Bed allocation in a public health care delivery system. *Management Science* 27 (5): 507–520.

Li, X. D., Beullens, P., Jones, D. and Tamiz, M. 2009. Optimal bed allocation in hospitals. *Lecture Notes in Economics and Mathematical Systems* 618: 253–265.

Meskens, N., Duvivier, D. and Hanset, A. 2013. Multi-objective operating room scheduling considering desiderata of the surgical team. *Decision Support Systems* 55: 650–659.

Nguyen, J., Six, P., Antonioli, D., Glemain, P., Potel, G., Lombrail, P. and Le Beux, P. 2005. A simple method to optimize hospital beds capacity. *International Journal of Medical Informatics* 74 (1): 39–49.

Tang, J., Yan, C. and Cao, P. 2014. Appointment scheduling algorithm considering routine and urgent patients. *Expert Systems with Applications* 42: 4529–4541.

Wang, S., Ma, Q. and Guan, Z. 2009. A fuzzy shifting bottleneck procedure for patient scheduling. *Proceedings of 2009 IEEE International Conference on Grey Systems and Intelligent Services*, November 10–12, Nanjing, China, 1566–1569.

Appendix: Fuzzy Set Theory Concepts

Support: The support of a fuzzy set A is the crisp subset of X that contains all the elements of X that have a nonzero membership grade in A. Thus, support is defined in form

$$\text{supp } A = \{x \mid \mu_A(x) > 0 \text{ and } x \in X\} \qquad \text{(A1.1)}$$

α-level set (α-cut): The α-level cut of a fuzzy set A is a crisp set A_α that contains all the elements of the universal set X with a membership grade in A greater than or equal to a prespecified α value. Therefore,

$$A_\alpha = \{x \mid \mu_A(x) \geq \alpha \text{ and } x \in X\} \qquad \text{(A1.2)}$$

Extension principle: The extension provides a general extension of the crisp mathematical concepts to fuzzy the fuzzy environment. Let f be a mapping from X to Y and A be a fuzzy set of X defined by $A = \sum_i \mu_A(x_i)/x_i$. Then, the extension principle states that

$$
\begin{aligned}
f(A) &= f(\mu_A(x_1)/x_1 + \mu_A(x_2)/x_2 \ldots + \mu_A(x_n)/x_n) \\
&= \mu_A(x_1)/f(x_1) + \mu_A(x_2)/f(x_2) \ldots + \mu_A(x_n)/f(x_n)
\end{aligned}
\qquad \text{(A1.3)}
$$

Normality: A fuzzy set is said to be normalized if and only if the supremum of $\mu_A(x)$ over X is unity—that is, $\sup_x \mu_A(x) = 1$. It follows that a nonempty subnormal fuzzy set can be normalized by dividing each $\mu_A(x)$ term by the factor $\sup_x \mu_A(x)$. Therefore, a $\mu_A(x)$ is a real number in the interval [0,1] with the grades 0 and 1 representing nonmembership and full membership in a fuzzy set, respectively. In this study, a characteristic function is always assumed normalized.

Fuzzy decision modeling: Decision making under fuzziness stems from the following basic elements: a fuzzy goal G in X, a fuzzy constraint C in X, and a fuzzy decision D in X, where X is a nonfuzzy space of alternatives. The objective is to satisfy C and attain G, which leads to the fuzzy decision

$$\mu_D(x) = \mu_C(x) \wedge \mu_G(x)$$

This yields the "fitness" of an $x \in X$ as a solution to the decision-making problem. It is logical to define the solution with the highest degree of membership to the fuzzy decision set as the optimal nonfuzzy decision. Consequently, a logical choice is x^* ∈ X such that

$$\mu_D(x^*) = \sup_{x \in X} \mu_D(x) = \sup_{x \in X}\left(\mu_C(x) \wedge \mu_G(x)\right)$$

Fuzzy decision: Given n fuzzy goals $G_1,\ldots, G_n$ and m fuzzy constraints $C_1,\ldots, C_m$ in a space of alternatives (or variants) X. Then, for each alternative $x \in X$, the resultant decision D is the intersection or confluence of goals $G_1,\ldots, G_n$ and the given constraints $C_1,\ldots, C_m$. The fuzzy set D is defined by the membership function $\mu_D(x)$ as

$$\mu_D(x) = \mu_{G_1}(x) * \ldots * \mu_{G_n}(x) * \mu_{C_1}(x) * \ldots * \mu_{C_m}(x)$$

where "*" is an aggregation (confluence) operator.

It follows from this definition that the maximizing decision is defined as $x^{opt} \in X$ such that

$$\mu(x^{opt}) = \max_{x \in X} \mu_D(x)$$

In the case when $m = n = 1$, the relationship is depicted in Figure A1.1. Consider a fuzzy goal G and fuzzy constraints C in a space of alternatives X. Then, with a minimum aggregation operator "∧," G and C combine to form a decision D, which is a fuzzy set given by $D = G \wedge C$. In terms of optimization, a fuzzy decision serves the purpose of an objective function introducing an order in the space of alternatives, where the performance of the decision making is given by

$$\mu_D(x) = \min\{\mu_G(x), \mu_C(x)\} = \mu_G(x) \wedge \mu_C(x)$$

Note that the intersection of fuzzy sets is defined in the possibilistic sense by the min-operator.

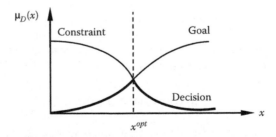

FIGURE A1.1 Decision as confluence of goal and constraint.

Index

Page numbers ending in "f" refer to figures. Page numbers ending in "t" refer to tables.